Rhetoric and the Early Royal Society

Scholarly Communication

VOLUME 3

The titles published in this series are listed at *brill.com/sc*

Rhetoric and the Early Royal Society

A Sourcebook

Edited by

Tina Skouen and Ryan J. Stark

BRILL

LEIDEN | BOSTON

Frontispiece to Thomas Sprat's *History of the Royal Society* (London, 1667). Copyright: The Royal Society. For information about the image, see Michael Hunter, *Science and Society in Restoration England* (Cambridge University Press, 1981), pp. 194–197.

Library of Congress Cataloging-in-Publication Data

Rhetoric and the Early Royal Society : a sourcebook / Edited by Tina Skouen and Ryan J. Stark.
pages cm. -- (Scholarly Communication; Volume 3)
Includes bibliographical references and index.
ISBN 978-90-04-28369-5 (hardback : alk. paper) 1. English language--Early modern, 1500-1700--Rhetoric. 2. English language--18th century--Rhetoric. 3. Royal Society (Great Britain)--History. 4. Literature and science--Great Britain--History. 5. English philology. I. Skouen, Tina, editor. II. Stark, Ryan J., editor.
PE1083.R38 2014
808'.04209032--dc23

2014034769

This publication has been typeset in the multilingual "Brill" typeface. With over 5,100 characters covering Latin, IPA, Greek, and Cyrillic, this typeface is especially suitable for use in the humanities.
For more information, please see www.brill.com/brill-typeface.

ISSN 1879-9027
ISBN 978-90-04-28369-5 (hardback)
ISBN 978-90-04-28370-1 (e-book)

This book is printed on acid-free paper.

Contents

List of Figures

Notes on Contributors

Frédérique Aït-Touati

is a Supernumerary Teaching Fellow at St John's College, Oxford, where she gives seminars and tutorials on French literature and thought, focusing on writing from the early modern period. She also holds an affiliate Professor position at Sciences-Po Paris. She is the author of *Fictions of the Cosmos. Science and Literature in the Seventeenth Century* (University of Chicago Press, 2011) and editor, with Anne Duprat, of *Histoires et savoirs. Anecdotes scientifiques et sérendipité aux XVIe et XVIIe siècles* (Peter Lang, 2012). Another book is forthcoming, with Stephen Gaukroger: *Le Monde en images. Voir, représenter, savoir, de Descartes à Leibniz* (Garnier).

Peter Dear

is Professor of History at Cornell University, where he specializes in the history of European science in the seventeenth century. His books include *Revolutionizing the Sciences. European Knowledge and Its Ambitions, 1500–1700* (Princeton University Press, second edition, 2009) and *The Intelligibility of Nature. How Science Makes Sense of the World* (University of Chicago Press, 2006).

Alan G. Gross

is Professor of Communication Studies at the University of Minnesota, Twin Cities. He teaches rhetorical theory, visual communication, and the rhetoric of science, and his books include *The Rhetoric of Science* (Harvard University Press, 1990 and 1996), and *Starring the Text. The Place of Rhetoric in Science Studies* (Southern Illinois University Press, 2006).

Joseph E. Harmon

is senior technical communicator at Argonne National Laboratory. Among his publications are *The Craft of Scientific Communication*, with Alan G. Gross (University of Chicago Press, 2010) and *Communicating Science. The Scientific Article from the 17th Century to the Present*, with Alan G. Gross and Michael Reidy (Oxford University Press, 2002).

Richard Nate

is Professor of English Literature at the Catholic University of Eichstätt-Ingolstadt. He teaches science, rhetoric, and literature from the early modern period to the present, among other subjects, and his most recent books include *Wissenschaft, Rhetorik und Literatur. Historische Perspektiven* (Königshausen & Neumann, 2009) and *Biologismus und Kulturkritik. Eugenische Diskurse der Moderne* (Königshausen & Neumann, 2014).

Michael S. Reidy
is Associate Professor of History at Montana State University, where he teaches the history of science and technology and the history of mountaineering, among other topics. His books include *Tides of History. Ocean Science and Her Majesty's Navy* (University of Chicago Press, 2008) and *Communicating Science. The Scientific Article from the 17th Century to the Present*, with Alan G. Gross and Joseph E. Harmon (Oxford University Press, 2002).

Tina Skouen
is Associate Professor of English Literature at the University of Oslo. Her publications include "Margaret Cavendish and the Stigma of Haste," *Studies in Philology* 111,3 (2014) and "The Rhetoric of Passion in John Donne's Holy Sonnets," *Rhetorica: A Journal of the History of Rhetoric* 27, 2 (2009).

Ryan J. Stark
is Associate Professor of English and Philosophy at Corban University. His interests include rhetorical theory, satire, and early modern literature and science. He is the author of *Rhetoric, Science, and Magic in Seventeenth-Century England* (Catholic University of America Press, 2009) and several articles on Renaissance and early Enlightenment thought.

Robert E. Stillman
is Professor of English at the University of Tennessee. He specializes in early modern literature and the rhetoric of science, and his publications include *The New Philosophy and Universal Languages in Seventeenth-Century England. Bacon, Hobbes, and Wilkins* (Bucknell University Press, 1995) and *Sidney's Poetic Justice. The Old Arcadia, Its Eclogues, and Renaissance Pastoral Traditions* (Bucknell University Press, 1986).

Michael Wintroub
is Associate Professor of Rhetoric at The University of California, Berkeley. Among his interests are scientific revolutions, humanism and court culture, and epistemology. He is the author of *A Savage Mirror. Power, Identity and Knowledge in Early Modern France* (Stanford University Press, 2006) and several articles including an award-winning study discussing the status of practical mathematics in the early modern period, "The Heavens Inscribed. The Instrumental Poetry of the Virgin in Early Modern France," *British Journal for the History of Science* 42, 2 (2009).

Sources

All the articles in this volume, with the exception of the Introduction, have been previously published.

1. "*Totius in verba*. Rhetoric and Authority in the Early Royal Society," by Peter Dear, was published in *Isis* (University of Chicago Press) 76, 2 (1985), 144–161. Used with permission.
2. "Rhetoric in the Early Royal Society," by Richard Nate, was published in *Rhetorica Movet. Studies in Historical and Modern Rhetoric in Honour of Heinrich F. Plett*, eds. Thomas O. Sloane and Peter L. Oesterreich (Brill, 1999), 215–231.
3. "Language Reform in the Late Seventeenth Century," by Ryan J. Stark, was published in *Rhetoric, Science and Magic in Seventeenth-Century England* (Catholic University of America Press, 2009), 47–87. Used with permission.
4. "Argument and 17th-Century Science. A Rhetorical Analysis with Sociological Implications," by Alan G. Gross, Joseph E. Harmon and Michael S. Reidy, was published in *Social Studies of Science* (Sage Publications Ltd) 30, 3 (2000), 371–396. Used with permission.
5. "Invitation and Engagement. Ideology and Wilkins's Philosophical Language," by Robert E. Stillman, was published in *Configurations* (Johns Hopkins University Press) 3, 1 (1995), 1–26. Used with permission.
6. "'The Spirit of Invention.' Hooke's Poetics for a New Science in *An Attempt to Prove the Motion of the Earth by Observation*," by Frédérique Aït-Touati, was published in *Etudes Epistémè* (Etudes-episteme.org; Institut du Monde anglophone, Sorbonne Nouvelle) 14 (2008), 105–121. Used with permission.
7. "The Looking Glass of Facts. Collecting, Rhetoric and Citing the Self in the Experimental Natural Philosophy of Robert Boyle," by Michael Wintroub, was published in *History of Science* (Science History Publications Ltd) 35 (1997), 189–217. Used with permission.
8. "Science versus Rhetoric? Sprat's *History of the Royal Society* Reconsidered," by Tina Skouen, was published in *Rhetorica. A Journal of the History of Rhetoric* (University of California Press) 29, 1 (2011), 23–52. Used with permission.

Introduction

Tina Skouen and Ryan J. Stark

In the last twenty years, scholars from several countries and in numerous disciplines have produced work on rhetoric and the early Royal Society (est. 1660). Despite wide geographical and multidisciplinary appeal, however, there is no single collection on the subject, and professors and graduate students will find material scattered in journals, monographs, and anthologies, and through many different online portals, if found at all. In response to this lacuna, we have created a sourcebook, a touchstone text for those interested in how the early Royal Society profoundly shaped modern scholarly communication.

Our aim in this introduction is to evaluate the early Society's pioneering role in reimagining the nature of facts, words, print, authority, and audience, among other things, which ultimately gave rise to a new species of writing and reading: the modern scholarly article and its concomitant disinterested, objective reader. We also examine the early Society's related commitment to performance and display, especially through the personal witnessing of live demonstrations, a practice that continues today—most commonly via digital technology. Finally, we clarify the Society's rhetorical practices by juxtaposing them against some of the competing rhetorical practices at the time, including those of the theosophists and the free thinkers.

Our introduction is followed by a collection of reprinted essays spanning a broad range of leading international journals and academic outlets. We have divided this collection into two parts, the first of which highlights key issues that help us to understand how the English natural philosophers defined the language of science. The opening three articles discuss the natural philosophers' partiality towards empirical evidence, as opposed to ancient authority, and their famous and controversial preference for things, not words. Moreover, these articles explore the philosophers' desire to represent in a plain style the empirical evidence discovered by the new methods of experimentalism. These issues are reviewed from the perspectives of, respectively, classical rhetoric, early modern science, and religion (see the chapters by Dear, Nate and Stark). The final article in this section offers a comparative analysis of the nature of scientific writing in early modern England and France (Gross, Harmon, Reidy), paying special attention to the topic of rhetorical probability and certainty. The second part of the collection presents four case studies of founding members of the Royal Society, focusing especially on their contributions to rhetorical and lingustic theory. The first paper meticulously examines John Wilkins's

 | DOI 10.1163/9789004283701_002

(1614–1672) attempt to develop a universal language (Stillman), perhaps the most ambitious and—arguably—naïve linguistic effort of the early modern period. This is followed by two ground-breaking arguments concerning, respectively, Robert Hooke's (1635–1703) "poetics of science" (Aït-Touati) and Robert Boyle's (1627–1691) adaptation of the rhetorical concept of *enargeia* (Wintroub). The volume's final essay concentrates on the way in which the Royal Society's chosen apologist, Thomas Sprat (*bap.* 1635, *d.* 1713), sought to reinvent a Ciceronian commonplace for use in modern society (Skouen). At the end of this introduction, we will give further details on the contents of each of the contributions.

Science, Rhetoric and Book History

Any discussion of rhetoric and the Royal Society should start with a recognition that the Society has at times "eschewed rhetoric," as Charles Bazerman writes in the Routledge *Encyclopedia of Rhetoric and Composition*.[1] Spurred on by Francis Bacon's (1561–1626) dream of linguistic transparency, the early Royal Society was obsessed with how to read the Book of Nature correctly.[2] The Society's motto, *nullius in verba* (take nobody's word for it), reflected the experimentalists' conviction that the search for knowledge should be conducted not through textual studies but rather through the direct, plain observation of the natural world. In their endeavor to avoid misrepresentation, members of the Royal Society considered (but hardly succeeded in) purging their language of rhetorical tropes and figures. The Royal Society's statutes of 1728, for example, contain the following article concerning written communication: "In all Reports of Experiments to be brought into the Society, the Matter of Fact shall be barely stated, without any Prefaces, Apologies, or Rhetorical Flourishes, and entered so into the Register-Book, by order of the Society."[3] This article shows that the Royal Society's concern with rhetoric and style was both real and official. Yet scholars have disagreed on the wider implications of the Society's rigid statements about language, as were expressed in numerous publications since the Society's foundation in 1660. According to Thomas Sprat's *History of the Royal Society* (1667), the members of the Society advocated a "mathematical plainness."[4]

1 *Encyclopedia of Rhetoric and Composition*, ed. T. Enos, 645.

2 D.R. Olson, *The World on Paper*, 160–178. See also Dear, in this volume.

3 The quotation from Chapter V, Article IV, of the statutes is cited by R.F. Jones, "Science and English Prose Style in the Third Quarter of the Seventeenth Century," 84.

4 T. Sprat, *The History*, 113.

But did this mean that they put a ban on rhetoric in all contexts? Or did the prescribed plain style only apply to scientific discourse, as suggested by the statutes of 1728? There is a long tradition among historians of rhetoric for placing at least some of the responsibility for rhetoric's supposed demise in the early modern period on the Royal Society (the essential secondary literature is reviewed in the contributions by Nate, Stark and Skouen in this volume). If nothing else, one can easily agree with Jennifer Richards's observation about Thomas Sprat's attitude toward the ancient art of rhetoric: "He helps us understand our own popular dismissal of this art."[5]

The Society's explicit goal was to reform the language of natural philosophy in particular, not of prose writing or literature in general. This point was firmly asserted by Brian Vickers in a paper that was first read at the William Andrews Clark Library in 1980, and then published as a 73-page essay in 1985 ("The Royal Society and English Prose Style. A Reassessment"). Vickers responded to a longstanding debate that began with Richard Foster Jones (*The Seventeenth Century. Studies in the History of English Thought and Literature from Bacon to Pope*, 1951), Wilbur S. Howell (*Logic and Rhetoric in England, 1500–1700*, 1956), and Morris W. Croll (*"Attic" and Baroque Prose Style. The Anti-Ciceronian Movement*, 1969). This debate was focused primarily on sentence-level issues and a confining concept of rhetoric (the debate is surveyed by Nate and Stark). Vickers's most significant claim, perhaps, was that one cannot trust the seventeenth-century scientists' accounts of style to be "neutral or accurately descriptive"; instead the debates generally resulted from "animus, or controversy, or party politics, or religious dispute."[6] In other words, the textual evidence ought to be checked against its historical and cultural contexts, a scholarly practice that Vickers found lacking in the previous criticism.[7] Since the 1980s, such awareness of context has informed the emergence of New Historicist criticism in literary studies and the development of two additional and related research fields that are especially relevant to the study of rhetoric and the early Royal Society, namely the rhetoric of science and the history of the book.[8]

The Royal Society's concern with appropriate language and style was also motivated by its consideration of the audience for science. This point may be illustrated by reference to Robert Boyle's thoughts on the style of the experimental essay in 1661: "certainly in these Discourses, where our Designe is only

---

5 J. Richards, *Rhetoric*, 6.

6 B. Vickers, "The Royal Society and English Prose Style," 23.

7 *Ibid.*, 15.

8 For a review of recent developments in rhetoric of science, see the opening chapter of A.G. Gross, *Starring the Text*, 3–19. See also R. Darnton, "What Is the History of Books?"

to inform Readers, not to delight or perswade them, Perspicuity ought to be esteem'd at least one of the best Qualifications of a style."[9] Today's readers are likely to doubt Boyle's claim that the experimental reports were initially designed "only to inform." "What is science without objections and desires?" asked Charles Bazerman rhetorically in his 1988 study of the experimental article as a genre.[10] Science is not just a matter of testing and re-testing for the sake of finding the truth, but it is also about laying claims on the truth. The rise of rhetoric of science as a research field has led to a greater awareness that language ought to be considered as more than a tool by which one communicates scientific knowledge. As Alan G. Gross has convincingly argued, language can also be constitutive of such knowledge,[11] a point Jeanne Fahnestock also advances in *Rhetorical Figures in Science* (1999), which offers ample evidence that metaphor, antithesis and various other figures have functioned as resources for scientific invention and reasoning.[12]

One could argue more broadly, too, that the making of the new sciences and of scientific publications were two sides of the same coin. In *Plus ultra, or, The Progress and Advancement of Knowledge since the Days of Aristotle* (1668), Joseph Glanvill (1636–1680), for example, reasons that there are two paths for the advancement of knowledge. One is by studying nature, and the other is "By *improving INTERCOURSE* and *COMMUNICATIONS*."[13] In describing the latter path, Glanvill explains that the Royal Society has sought to increase communication by personal interaction (sometimes involving overseas travels) and through the circulation of texts.[14] Glanvill devotes special attention to the invention of printing as an unparalleled means by which one can preserve, disseminate and further improve learning. Before print, "*Copies* of excellent things could not be so much *dispersed*, nor so well *preserv'd* either from the *Corruptions*

9 Quoted from the essay serving as a preface to *Certain Physiological Essays Written at Distant Times, and on Several Occasions* (1661), as cited by J.E. Harmon and A.G. Gross (eds.), *The Scientific Literature*, 36.

10 C. Bazerman, *Shaping Written Knowledge*, 79. See also G. Cantor, "The Rhetoric of Experiment," 161.

11 See Gross, *Starring the Text*, 5, 14–15.

12 J. Fahnestock, *Rhetorical Figures in Science*, esp. 23–24.

13 *Plus ultra*, 9. For Glanvill's views on language, see the essays by Nate and Stark in this volume.

14 Glanvill considers both the art of printing and the art of navigation by the compass in *Plus ultra*, 77–80. He does not mention the improvements in postal services, which was the third factor contributing to "the fundamental changes in communications in the early modern period" according to S.J. Harris, "Networks of Travel, Correspondence, and Exchange," 347.

of *Time* or *Design*. The *Charge* of *Books* was very *great, forgeries frequent*, and *mistakes* of *Transcribers numerous*. They were quickly swept away out of those few *Libraries* in which they were, by *Fire* and *Violence*, or *spoiled* by *Dust* and *Rottenness*."[15] Speaking on behalf of the new sciences, Glanvill pronounces that "we have *Advantages* above *Aristotle*, and, which is much more, *above* all elder Times, for *mutual Communications*, and *impartments* of our *Notices, Observations, Experiments*, and *Performances* for the *increase* of *Science*."[16]

Glanvill thus testifies to the importance which the early Royal Society placed on printed material, despite the Society's insistence that knowledge ought to be based on the study not of printed books, but of the Book of Nature.[17] The Society's intention of "imparting" the reports of their various findings to an international scholarly community manifested itself most clearly in the 1665 launch of the *Philosophical Transactions*, commonly considered the first science journal.[18] Glanvill's belief in the fixity and incorruptibility of printed texts and his linking of the "increase" of science with the rise of print were later to become commonplace beliefs, although, as we shall see, in recent years these commonplaces have been contested. In *The Printing Press as an Agent of Change* (1979), Elizabeth L. Eisenstein famously made the case that the rise of science was closely related to the rise of print: because of the rapid spread of printed collections documenting scientific observations, individual scholars such as Nicolaus Copernicus (*d.* 1543) and Tycho Brahe (*d.* 1601) were able to check their findings against those of others, whether past or present, thus making significant progress.[19] Similar arguments concerning the relation between text and knowledge production have also been forwarded in literacy studies. According to Jack Goody, the invention of writing around 3000 BC made it possible to "make, record and hence *compare* repeated observations in a precise way."[20] Before the Greeks, the Babylonians "had already taken us along the road

15 *Plus ultra*, 78.

16 *Ibid.*, 77. In the seventeenth century, "science" (Lat. *scientia*) was not a term denoting studies of nature (i.e. natural sciences) but it could be used to designate "any body of properly constituted knowledge," as S. Shapin observes in *The Scientific Revolution* 5, n. 6. The term "scientist" had not yet been invented.

17 See A. Johns, "Science and the book," esp. 275, 298; A. Johns, *The Nature of the Book*, esp. 45.

18 Gross and Harmon, *The Scientific Literature*, 39. For an overview of the early *Philosophical Transactions*, see *ibid.*, 1–5; M. Hunter, *Establishing the New Science*, 255. For an in-depth analysis of the contents of the *Transactions*, see Gross, Harmon, Reidy in this volume.

19 See Chapter 7 of *The Printing Revolution in Early Modern Europe* (esp. 231–238), where Eisenstein summarizes an argument that she first raised in *The Printing Press as an Agent of Change*.

20 Goody, *The Interface between the Written and the Oral*, 75–76.

to 'science,' and this journey," Goody argues, "was intrinsically linked to their ability to communicate in the written channel."[21] The established view, as voiced by Joseph Glanvill in the seventeenth century, is therefore that the advancement of knowledge and printed communication go hand in hand. "Writing…serves to separate and distance the knower and the known and thus to establish objectivity," writes Walter Ong in *Orality and Literacy* (first publ. 1982).[22]

With a printed text, there is also a claim to accuracy, to finality. According to Ong, the introduction of print meant that old debates and controversies could be brought to an end, while at the same time ensuring that scientific observations could be accurately represented in any number of "exactly repeatable" copies.[23] More recent research in the history of the book, however, has shown such understandings of the effects of print to be both anachronistic and overly simplistic. As Adrian Johns has argued in *The Nature of the Book*, today's notion of "uniform editions, mass reproduction, and typographical fixity" must be checked against the actual practices of book production throughout the hand-press period (1500–1800).[24] A key insight in the work of D.F. McKenzie and others has been that printed books were the result of complex "working processes" involving many different crafts and interests running alongside those of the author.[25] Battles were fought over the printing of and rights to the material. In the seventeenth century, authors did not yet own the copyright to their work, nor did they have much say in the publication process. The book trade rather belonged to booksellers and printers.[26] At the printing house, even the most distinguished scholar-nobleman would be at the commands of the master printer, who did not always have the means or will to prevent an incompetent or overly creative setting of a complex mathematical table.[27] Complaints

21 *Ibid.*, 76. See also J. Goody and I. Watt, "The Consequences of Literacy."

22 Ong, *Orality and Literacy*, 112. See also 14–15.

23 *Ibid.*, 125, 129–133.

24 Johns, *Nature of the Book*, 28. The rough dating of the hand press period is from Darnton, "What Is the History of Books?" 19.

25 D.F. McKenzie, *Bibliography and the Sociology of Texts*, 3. See also R. Chartier, *The Order of Books*, esp. 9–10; Johns, *Nature of the Book*, esp. 2–5; E.L. Eisenstein, *The Printing Press as an Agent of Change*, 55–56.

26 J. Raven, *The Business of Books*, 4–6; J. Feather, "The British Book Market 1600–1800"; A. Weedon, "The Economics of Print." For details on the Copyright Act of 1709, "The World's Earliest Copyright Statute," see the entry on "An Act for the Encouragement of Learning," in *The Oxford Companion to the Book*, eds. M.F. Suarez and H.R. Woudhuysen, vol. 1, 448.

27 Johns, "Science and the Book," 290–292. See also Johns, *Nature of the Book*, 104. Because of the difficulties involved in reproducing tables and diagrams, "manuscripts would have an

about printing faults flourish in works of all genres, and prefaces and errata sheets read as records of all that could go wrong in the process. And much did. Indeed, writers in all genres were subject to typographic misrepresentation, piracy, and unauthorized publication, but the consequences were potentially bigger if a scientific author "could not guarantee the fidelity" of the texts and figures presented.[28] Consequently, the Royal Charter of 1662 gave the Society the right to appoint its own printers and engravers and to license and publish learned works in direct competition with the Stationers' Company, the book trade organization, otherwise exercising monopoly in London and, in effect, the whole of England.[29] Similar privileges had also been granted to the universities of Cambridge (in 1534) and Oxford (in 1586), but not until the end of the seventeenth century did the scholarly community fully compete with the Stationers' Company in terms of asserting authority and making a profit.[30]

For the Royal Society's first secretary, Henry Oldenburg (*c.*1619–1677), it was also a matter of great concern to prevent foreign reprints and translations of the *Philosophical Transactions*, because this could potentially undermine the Society's as well as the individual contributor's claim to the findings presented.[31] Adrian Johns makes the point that Oldenburg was faced with the dilemma of how to spread the word about the Society's progress, without risking that it lose control of the scholarship being produced.[32] Thus, when the booksellers John Martyn and James Allestry were accredited as Printers to the Royal Society in 1663, they had to swear an oath that they would never issue reprints or translations unless licensed by the Society.[33] From Oldenburg's

---

important continuing place in scientific discourse long into the age of print," writes A. Pettegree in *The Book in the Renaissance*, 279.

28 Johns, "Science and the Book," 290. Consider, for example, the problem of producing accurate, large-scale maps for use in navigation, as discussed in Olson, *World on Paper*, 203–212; Johns, *Nature of the Book*, esp. 452. The problems of piracy in the early modern age are analyzed in great detail by A. Johns, *Piracy*, 41–143.

29 Johns, *Nature of the Book*, 491–494. See also the translation of the *Charta Prima*, the first Royal Charter, in *The Record of the Royal Society*, 234–235. For an overview of the licensing system and other forms of regulation of the book market in the period from 1557 (when the Company of Stationers was chartered) to the post-Restoration, see J. Feather, *A History of British Publishing*, 29–50.

30 A. Murphy, "The History of the Book in Britain, *c.* 1475–1800," 174; Johns, "Science and the Book," 291–295; D. McKitterick, "University Printing at Oxford and Cambridge."

31 Johns, *Nature of the Book*, 514–521.

32 *Ibid.*, 515.

33 *Ibid.*, 493; C.A. Rivington, "Early Printers to the Royal Society, 1663–1708," 2–3.

standpoint, the Society's authorization gave a sense of closure and legitimacy to the printed page, but such fixity of print—real or imagined—did not establish by itself a rhetoric of objectivity. This is an important point. As Johns, Olson, Stephen B. Dobranski, and others have shown, science writing, like all other Renaissance and early modern literature, depended also on the active reader to complete and validate its meaning.[34] Whereas the author could be duly accredited for the work thus presented in print, and the page or printing technology could be duly credited for vividly preserving the experiments, the project of building objective truth continued beyond the authors and the printed pages, when the text was delivered finally to a qualified and empowered readership. Scientific objectivity, that is, was a dynamic partnership from the start, a collaboration between an authorizing agent and a validating literate public, culminating in what has come to be known as modern scholarly communication.

Whereas the Society's "aggressive intervention in the realm of print" remains the main focus of Adrian Johns's pivotal study, his perspective also extends to the practices of oral performance and conversation that help to define the new science.[35] Most notably, Johns observes a certain convergence between the emerging coffeehouse culture and experimentalism.[36] In the words of John Wilkins, one of the Society's founders, the scientific endeavor in fact depended on "mutual converse" and a "quality of sociableness" rather than mere "bookish" studies.[37] The coffeehouse, like the Society's lodgings in Gresham College and later Arundel House, was a place which thrived on both the spoken and printed word and where civility was put to the test through sometimes heated discussions.[38] Wary of controversy and debate, the Royal Society regarded scholastic dispute as unproductive and political conflict as downright disruptive.[39]

34 Johns, *Nature of the Book*, 30–37; Olson, *World on Paper*, 173–178; S.B. Dobranski, *Readers and Authorship in Early Modern England*, 5–17, 29–34. See also Elizabeth Spiller, *Science, Reading, and Renaissance Literature*, 137–177. Readers' active engagement in annotating, correcting or completing material presented in the *Philosophical Transactions* and other printed sources contributes to undermine the strict distinction between manuscript versus print, as E. Yale has argued in "Making Lists" and "With Slips and Scraps."

35 Johns, *Nature of the Book*, 44.

36 *Ibid.*, esp. 67–68, 553–560; A. Johns, "Coffeehouses and Print Shops." See also L. Stewart, *The Rise of Public Science*, esp. 143–146.

37 Cited by Johns, *Nature of the Book*, 470. For more on Wilkins, see Stillman in this volume.

38 See e.g. the account of the prolonged public dispute between the astronomer John Flamsteed and Robert Hooke, in Johns, *Nature of the Book*, 551–560.

39 S. Gaukroger, *The Emergence of a Scientific Culture*, 40, formulated as a critique of Toby E. Huff's argument in *The Rise of Early Modern Science* that early modern science

As Glanvill explained in *Plus ultra*, "The *Modern Experimenters*" were inclined to think that "the *Philosophers* of elder Times" took the wrong path when they engaged in "a *Labyrinth* of *Talk*."[40] As Stephen Gaukroger and others have maintained, the Fellows therefore sought to resolve scholarly disagreements through a "shared witnessing of experiments" rather than 'adversarial argument."[41] Those who were called upon to witness the performance of an experiment were expected to vouch for the result in case of outside opposition, not only from different schools of thought or competing scientific bodies, but also from the public sphere.[42]

As public institutions, the Royal Society and the coffeehouse contribute equally to "the discursive construction of modernity" according to Richard Bauman and Charles L. Briggs.[43] Yet, one could argue that the early Royal Society engaged in a Janus-faced relationship with the general audience, on the one hand presupposing and seeking to interact with a Habermasian public sphere, on the other hand recoiling from the hubbub and noise of that sphere. As Bauman and Briggs have claimed, Habermas in his discussion of John Locke's (1632–1704) concept of "disinterested" or "dispassionate" speech might not have realized the extent to which such generalizations may have served to exclude the greater majority, and thus to legitimize social inequality.[44] Thomas Hobbes (1588–1679), whose "anti-experimentalism"—according to Steven Shapin and Simon Schaffer—prevented him from joining, may have been the first to observe that the site of experimentalism was in fact "a public space with restricted access."[45] Fuelled by Boyle's account in his *New Experiments Physico-mathematical* (1660) of the "fifty men of philosophy" meeting each week at Gresham College, Hobbes protests: "Why do you speak of fifty men? Cannot

---

thrived in "an adversarial culture." See also S. Schaffer and S. Shapin, *Leviathan and the Air-Pump*, 283, 341; R. Bauman and C.L. Briggs, *Voices of Modernity*, 27; A. Potkay, *The Fate of Eloquence in the Age of Hume*, 51–58. The elevated status of the witness is an important topic in several of the articles included in this volume, see esp. Dear; Gross, Harmon, Reidy; and the related discussion of the effects of vivid description (*enargeia*) in Wintroub.

40 Glanvill, *Plus ultra*, 7.

41 S. Gaukroger, *Francis Bacon and the Transformation of Early-Modern Philosophy*, 223–224.

42 Johns, *Nature of the Book*, 472. See also Bazerman, *Shaping Written Knowledge*, 140–141.

43 Bauman and Briggs, *Voices of Modernity*, 2. They refer (*ibid.*, 194 and *passim*) to the concept of "the rational public sphere," as developed by Jürgen Habermas in *The Structural Transformation of the Public Sphere*. For the Royal Society as a chartered, public institution, see also Hunter, *Establishing the New Science*, 2.

44 Bauman and Briggs, *Voices of Modernity*, 194, 63.

45 Shapin and Schaffer, *Leviathan and the Air-Pump*, 139, 336.

everyone who wishes come, …and give his opinion on the experiments which are seen, as well as they?' "Not at all," responds his Boylean interlocutor: "the place where they meet is not public."[46] Hobbes does not seem to have been bothered by the Society's rhetoric of presence; in fact, the idea in principle appealed to him. For example, he praised the ancient Greek historian Thucydides for the historian's ability—through perspicacious rhetoric—"to make his auditor a spectator."[47] What Hobbes objected to was the Society's practice of excluding gadflies like himself. Hobbes's attitude was the exact opposite of that expressed by Groucho Marx, who famously refused to join any club that would accept him as a member. Nonetheless, by making the Royal Society appear as a closed circle of "self-selected" witnesses, Hobbes reduced the Society's claims of the truth to myth, their methods appearing as merely accidental: "First, experiments are produced, and then on another day, whatever the cause of the phenomenon is suspected to be, someone orally explains it, if he can. For [they] do not have enough trust in written natural histories."[48] From Hobbes's ironic perspective, the performance of the new science consequently represents the triumph not of print and objectivity, but rather of situated, partial, oral communication.

The Rise of the Experimental Article

To the extent that one may consider both writing and knowledge-making as social practices or processes, the role of communicating in person can hardly be underestimated.[49] As is still the case today, academic research tends to develop through an integral process involving reading papers aloud and getting feedback from live audiences, engaging in discussions face-to-face or by letters (in this day and age, mostly electronic), and exchanging ideas in

46 Quoted from S. Schaffer's translation of Hobbes's Physical Dialogue, *Dialogus Physicus* (1661), *ibid.*, in the Appendix, 349–391 (350).

47 J. Rothkamm, *Institutio oratoria. Bacon, Descartes, Hobbes, Spinoza*, 161.

48 Schaffer's translation of Hobbes's Physical Dialogue in Shapin and Schaffer, *Leviathan and the Air-Pump*, 351 and see *ibid.*, 113.

49 For various understandings of science as a social practice, see eg. R.K. Merton, *The Sociology of Science*, w. ref. to the Royal Society *passim*; M. Hunter, *Science and Society in Restoration England*; Shapin and Schaffer, *Leviathan and the Air-Pump*; S. Shapin, *A Social History of Truth*; B. Latour, *Science in Action*, 21–62; Bazerman, *Shaping Written Knowledge*; C. Bazerman, "The Writing of Social Organization and the Literate Situating of Cognition. Extending Goody's Social Implications of Writing," esp. 215; Gross, *The Rhetoric of Science*, 163–179; Johns, *Nature of the Book*, 40–48.

laboratories or other collegial settings.[50] Yet, whereas today the distinction between a letter sent by (e)mail and a peer-reviewed journal article is very clear in terms of both presentation and accreditation, in the seventeenth century such distinctions were still in the making. Bazerman has emphasized that the experimental article as a genre developed from the exchange of notes and letters between individuals across Europe.[51] At least since the sixteenth century, scholars had engaged in epistolary networks exchanging letters containing observations and discoveries of the natural world, and sometimes specimens from plants or animals.[52] While a letter could easily be forwarded, transcribed or paraphrased, there were no set practices for collecting the many scattered observations contained in the letters and distributing them to a wider community of readers sharing the same interests. Commonly regarded as the head of a communications or correspondence network reaching across Europe, the Royal Society's secretary Henry Oldenburg turned the reading of such letters into an event, by reporting on his various correspondence at the Royal Society's meetings.[53] When he first launched the *Philosophical Transactions* in 1665, his idea was to share the contents of these letters with even more members of the learned community, in other words to make them public, thereby promoting the new science.[54] This practice in turn changed the way in which these letters were written.[55] Perhaps the best example is the one given by Alan G. Gross, Joseph E. Harmon and Michael Reidy in *Communicating Science*, namely Isaac Newton's (1642–1727) first paper on optics ("the first major scientific article in English"), published in the *Transactions* in 1672.[56] Newton wrote the paper in the form of a letter addressed to Henry Oldenburg, the journal's editor, "*in order to be communicated to the* R. Society," but his

---

50 Cf. B. Latour and S. Woolgar, *Laboratory Life*, 154–174.

51 Bazerman, *Shaping Written Knowledge*, 128–133.

52 Harris, "Networks of Travel, Correspondence, and Exchange," 344, with reference to P. Findlen, *Possessing Nature*, see esp. 157–158, 173.

53 See Gross, Harmon, Reidy in this volume. See also Hunter, *Science and Society*, 51–54; Hunter, *Establishing the New Science*, 245–256; Johns, *Nature of the Book*, 497–501; 514–515; Bazerman, "The Writing of Social Organization," 227–228; Gross and Harmon, *The Scientific Literature*, 2–3. For a broader discussion of scribal communities and manuscript circulation in general, see H. Love, *The Culture and Commerce of Texts*, with reference to the Royal Society's secretary, Oldenburg, on p. 147 and 182.

54 Hunter, *Establishing the New Science*, 248; Johns, *Nature of the Book*, 499.

55 Bazerman, *Shaping Written Knowledge*, 131–133.

56 Gross, Harmon and Reidy, *Communicating Science*, 68, and cf. the same authors' contribution to the present volume. See also Dear in this volume, discussing Newton's article from a different angle.

persuasive appeal to "the Reader" towards the end of the paper shows that he was quite aware that it would not only be read aloud at the Society's next meeting, but would also reach a much wider audience through publication in Oldenburg's journal.[57] To appear in print was, to be sure, a way of sharing ideas and making them public, but additionally it represented a solution to the problem of how to share information with others while at the same time securing the rights to one's findings.[58] Gross points to the paradox, first identified by Robert K. Merton and Harriet Zuckerman in 1971, that although the scientific enterprise is built on cooperation there is also bound to be competition between individuals striving for recognition.[59] In the mid-seventeenth century, a letter, lecture or handwritten entry into the Society's register could be enough to establish priority rights, but towards the end of the seventeenth century the dated printing of papers came to be recognized as the definite means of securing such rights, a function which dated journal publication still has today.[60]

Given this important function of print publication, the role of oral communication in early modern science understandably has received comparatively little attention. There is also the widely held belief that the transitions from oral to written cultures and from script to print represent "a step forward for enlightenment, science, rationality and modernity."[61] Moreover, whatever knowledge or practice not documented in print tends to remain obscure, a fact which largely explains why we know so little about women's contributions to early modern science. The house of Lady Ranelagh, Robert Boyle's sister, for example, has been described as "a centre for the new science being promoted by Robert Boyle and the Royal Society," and yet, as Sarah Hutton observes in her short biography in the *Oxford Dictionary of National Biography*, we have little evidence of the extent to which Lady Ranelagh (1615–1691) participated in the investigations and experiments in Boyle's private laboratory.[62] Although one can tell from Lady Ranelagh's letters and various other sources that she

---

57 Gross, Harmon and Reidy, *Communicating Science*, 69. Quotations from Newton's letter are from the electronic text of The Newton Project.

58 See R. Merton and H. Zuckerman, "Patterns of Evaluation in Science," 465–467; Gross, *Rhetoric of Science*, 174–176, reprinted in the revised version of the same work, Gross, *Starring the Text*, 175–177.

59 Gross, *Starring the Text*, 165, with primary reference to Merton and Zuckerman, "Patterns of Evaluation in Science."

60 Gross, *Starring the Text*, 166; Johns, *Nature of the Book*, 501; R.N. Giere, "Naturalism," 218.

61 M. Lyons, *A History of Reading and Writing in the Western World*, 34, with reference to Elizabeth Eisenstein and the thesis of the "printing revolution."

62 S. Hutton, "Jones, Katherine, Viscountess Ranelagh (1615–1691)."

(like many other women in this period) administered medicine and treated members of her family, she left no published writings to document further the extent of her learning.[63] Whereas women were free to participate in the informal settings of the household laboratory and private salon, they were not encouraged to seek publication nor indeed to take part in public discourse, and Margaret Cavendish, Duchess of Newcastle (*c.* 1623–1673), was the first and the only Englishwoman of the period to publish works in natural philosophy in her own name.[64] Women also participated in the epistolary networks that dominated the exchange of ideas in the first half of the century, but they were discouraged from any formal participation in the emerging institution of science. Case in point: Margaret Cavendish's exclusion from the Royal Society.[65] A defender of rationalist deduction like her friend Thomas Hobbes (they would both have qualified as "Dogmatists" in the eyes of Thomas Sprat), the Duchess of Newcastle would hardly have felt at home at Gresham College.[66] As Steven Shapin and Simon Schaffer have demonstrated, Hobbes considered the Royal Society as a space not for universal enquiry, but rather only for self-referential game-playing; that is, Hobbes seriously questioned the Society's ambition towards engaging the wider community.[67]

---

63 *Ibid.* See also L. Hunter, "Sisters of the Royal Society. The Circle of Katherine Jones, Lady Ranelagh." For women's contributions in the field of medicine, see also L. Hunter and S. Hutton (eds.), *Women, Science and Medicine 1500–1700;* E. Spiller (ed.), *Seventeenth-Century English Recipe Books.* The distinction between professional and domestic medicine generally "exemplifies the literate/non-literate division between the sexes," according to S.H. Mendelson, *The Mental World of Stuart Women*, 8. It should be noted, however, that even in literate households, much of the medical knowledge was transmitted by word of mouth, and that most of the medical books published in England in the sixteenth and seventeenth centuries were written in the vernacular and not in Latin, see A. Wear, *Knowledge and Practice in English Medicine, 1550–1680*, 40, 43, 50. For an argument that the pursuit of natural knowledge in early modern Europe was to a large extent "a *family* project," see A. Cooper, "Homes and Households," 225.

64 L.T. Sarasohn, *The Natural Philosophy of Margaret Cavendish*, 4. Cavendish was also "The First Englishwoman to Publish Extensively," according to N. Cotton, *Women Playwrights in English,* 44. For an overview of the general impact of women's writing in the period, see M.L. North, "Women, the Material Book and Early Printing."

65 Sarasohn, *The Natural Philosophy*, 5–6. For a review of Margaret Cavendish's legendary visit to the Royal Society on May 30, 1667, see *ibid.*, 25–33.

66 Compare Shapin and Schaffer, *Leviathan and the Air-Pump*, 136–139; and A. Battigelli, *Margaret Cavendish and the Exiles of the Mind*, 85–113. For Cavendish's repudiation of experimental natural philosophy see also Sarasohn, *The Natural Philosophy of Margaret Cavendish*, esp. 10–13; F. Aït-Touati, *Fictions of the Cosmos*, 174–190.

67 Shapin and Schaffer, *Leviathan and the Air-Pump*, 336.

Public Communication

Inspired by Francis Bacon, the Royal Society set itself the ambitious goal of making science an integral part of society.[68] Ideally, their methods of observation and validation were to be more or less the same as those used in everyday life, so that in principle almost any member of the lay public could participate.[69] As part of the same goal, it was deemed necessary to use an accessible language, in other words to employ such terms as were common among "Artizans, Countrymen, and Merchants" (as Thomas Sprat emphasizes) and, indeed, to prefer the vernacular over Latin.[70] The results of specific 'experimental events' were presented in writing not as isolated events but instead as 'facts,' a term which originally connoted human action but which came to be associated with proof of how nature behaves, whether on a universal scale (as Isaac Newton allowed) or under certain conditions (as with Robert Boyle's experiments with the air-pump).[71] What Boyle and his colleagues purported to do was to lay bare the "matters of fact" for everyone to see, and to bring every detail before the inner eye, so as to create a sense that the readers were actually present at the original performance of an experiment (an effect which Shapin and Schaffer term "virtual witnessing").[72] In so doing, they made nature their "strongest ally of persuasive argument," as Bazerman has emphasized.[73] The reputedly anti-rhetorical academy thus placed rhetoric at the very heart of the scientific endeavor, not merely because of its emphasis on performance and audience, but also because of its insistence on an open mode of

68 The Baconian principle of "openness" is stressed by Vickers, "The Royal Society and English Prose Style," 28, 35; Shapin, "Science and the Public," 995–996.

69 Shapin, "Science and the Public," 996.

70 T. Sprat, *The History*, 113. For the general preference of the vernacular, see Gross and Harmon, *The Scientific Literature*, xix. The difficulties involved in finding appropriate concepts in English are discussed in Vickers, "The Royal Society and English Prose Style," 28–29, 35–36.

71 P. Dear, "The Meanings of Experience," 124–129 (129). The etymology of "the fact" (Latin *factum* or deed) is also stressed in Latour and Woolgar, *Laboratory Life*, 174. See also R.W. Serjeantson, "Proof and Persuasion," 157–161, with reference to B. Shapiro, *The Culture of Fact* and L. Daston, "Strange Facts." Shapin and Schaffer devote considerable attention to "the means by which experiments can be said to produce matters of fact," in *Leviathan and the Air-Pump*, 3, and "matters of fact" *passim*. The chapter by Gross, Harmon and Reidy in the present volume focuses on how facts were generated and explained in seventeenth century science journals.

72 Shapin and Schaffer, *Leviathan and the Air-Pump*, 60–65; S. Shapin, *A Social History of Truth*, 179–180.

73 Bazerman, *Shaping Written Knowledge*, 321.

communication. Extending, as it were, their hands to the public, the natural philosophers chose the open palm of rhetoric over logic, traditionally conceived of as a closed fist.[74]

Studies of the market for public lectures and the advertisements for scientific books and instruments in the age of Newton have shown that there was a growing popular audience for science.[75] As Schaffer puts the matter, there is a strong element of spectacle and display in the practice of experimental natural philosophy in the eighteenth century.[76] The new science was advocated in numerous public venues, including notably the Ashmolean Museum at Oxford, opened in 1683 as a gift from Elias Ashmole, a founding member of the Royal Society.[77] A distinguished German scholar visiting in 1710 was shocked to find the place crowded with "all sorts of country-folk, men and women," and he was equally dismayed at the extent to which the Royal Society's in-house exhibitions in London were actually accessible to the lay public.[78] Yet, such elements of "public display" do not necessarily converge with the ideal of an actual "public presence" in the observation of nature and the validation of scientific knowledge.[79] In reality, there was much to distinguish everyday phenomena from scientific observations: as Lorraine Daston has pointed out, scientific observations were increasingly characterized by "special procedures, carried out by specially qualified people under special circumstances."[80] Daston refers

---

74 The image of a closed versus open fist is described in W.S. Howell, *Logic and Rhetoric in England, 1500–1700*, 3–4, 14–15. See also the section on Juan Luis Vives (1492–1540) in W. Rebhorn, *Renaissance Debates on Rhetoric*, 84–85.

75 M. Hunter, *Science and Society*, 84. Stewart, in *The Rise of Public Science* (esp. 101–141, 213–242) discusses the numerous popular lectures and demonstrations of experiments by Newtonian followers such as John Theophilus Desaguliers (1683–1744). Desaguliers is described as "the Royal Society's official experimenter" and Newton's "most illustrious popularizer" in R. Porter, *Enlightenment*, 142. For advertising, see J.R. Wigelsworth, *Selling Science in the Age of Newton*, 7. According to Wigelsworth (*ibid.*, 9, 13–14, 17), the advertisements in the *Philosophical Transactions* (and later, in newspapers) were increasingly aimed as much at the general public as to professional scholars and virtuosos, the purpose being partly to "create sales" and "grant legitimacy" to specific products and partly to "solicit public involvement in particular philosophical projects."

76 S. Schaffer, "Natural Philosophy and Public Spectacle."

77 Findlen, *Possessing Nature*, 147.

78 *Ibid.*, quoting from Zacharias Conrad von Uffenbach's (1683–1734) travel diary. See also *ibid.*, 148, and the discussion of the Royal Society's Repository in the article by Wintroub in the present volume.

79 Compare Schaffer, "Natural Philosophy and Public Spectacle," 1; and Shapin, "Science and the Public," 995.

80 L. Daston, "The Empire of Observation, 1699–1800," 91.

to the Paris Académie's *Histoire naturelle des animaux*, which was advertised in 1723 as containing "no facts that have not been verified by the whole Company, composed of people who have eyes for seeing these sorts of things, in contrast to the majority of the rest of the world."[81]

The sense that the practice of science was the privilege of an exclusive circle may have been even stronger with the Royal Society, which, unlike its French counterpart, consisted only of such men as were able to support the Society financially.[82] Whereas the Académie Royale has been compared to "a government-funded research institute," the Royal Society emerges as something of an élite "gentleman's club."[83] Despite the Society's stated aspirations towards engaging the entire society, the relationship with the public was precarious at best, and, to complicate matters further, much of the technical research conducted in the mathematical sciences—and in several other areas—was bound to be somewhat inaccessible to a wider audience anyway.[84] Henry Pemberton, Royal Society Fellow and editor of the third edition of Newton's *Principia*, identified the tension inherent in the new scientist's search for an audience: "Now, since Sir Isaac Newton's doctrine has been fully established by the universal approbation of all, who are qualified to understand the same; it is without doubt to be wished, that the whole of his improvements in philosophy might be universally known."[85] Pemberton's wish that Newton's philosophy be universally grasped is sharply mitigated by his awareness that science's validating readers must be those "who are qualified to understand the same."

But that the Society's Fellows expressed with some enthusiasm the goal of mass scientific literacy is itself a considerable achievement. We do not, for example, discover in the ancient world a desire for Galenic medicine to be understood by the masses, nor do we find in medieval or early Renaissance culture urgent calls for the poor to understand engineering, or optics. This leads us to an interesting question: From whom did the Royal Society inherit this vision of mass literacy, which—at least to a certain extent—presupposes the envisioning of an enormous educational system? Olson rightly contends

81 *Ibid.*, 93, and see 110, n. 56. The Parisian audience for science is outlined in A. Stroup, *A Company of Scientists*, 180–199. See also Gross, Harmon, Reidy in the present volume.

82 Shapin, *A Social History of Truth*, 123–124; Gaukroger, *The Emergence of a Scientific Culture*, 36, n. 93, with reference to Chapter 2 of M. Hunter, *The Royal Society and Its Fellows*. See also J. Gascoigne, "The Royal Society and the Emergence of Science as an Instrument of State Policy," 172.

83 Gross, Reidy, and Harmon, *Communicating Science*, 32; Gross and Harmon (eds.), *The Scientific Literature*, 42.

84 Shapin, "Science and the Public," 996.

85 H. Pemberton, *A View of Sir Isaac Newton's Philosophy*, 301.

that the Protestant Reformers gave the experimentalists the idea.[86] The argument goes as follows: pursuing Bacon's agenda, the Fellows posited an ideal audience capable of reading for themselves letters in the Book of Nature, written by God and readily accessible for all who know how to read properly, just as Luther and Melanchthon—in similar fashion during the Reformation—imagined a priesthood of lay believers capable of interpreting the Bible without the papacy's authoritarian decrees and the scholastic philosophers' gargantuan commentaries. The revolutionary idea of *sola scriptura* and the accompanying "literal" truths contained therein provides the sturdiest precedent for the new experimentalist's notion of Nature's Book and the concomitant "plain," self-evident or literal facts available to all, depending on how we might want to characterize the always-mysterious *things* in themselves. Herein lies the provocative connection between what Olson rather crudely deems "the Protestant way of reading Scripture" and the newly experimental way of reading Nature.[87] In both cases, religious and scientific, an empowered reader participates actively in a prescribed hermeneutical framework, and this audience-based hermeneutical capability, whether directed toward the Bible or Nature, is inherently rhetorical. In this way, Olson usefully shows that the Society's mantra of *nullius in verba* misleads in a certain way, because the thing self—in order to be understood as the thing itself—requires a sophisticated audience wielding a correct hermeneutic.[88] Put differently, Olson provides an attendant observation to the often-made point that the Royal Society's anti-rhetorical style is itself a powerful style (i.e., anti-rhetoric is rhetoric), namely that the Society's validating public sphere is an inherently rhetorical construct designed to collaborate with the new experimentalists in the production of facts. The validating public is not a passive *tabula rasa* upon which facts are to impress themselves; rather, they are active co-conspirators in the production of early modern knowledge. In the words of the old Native American medicine woman at Bertrand Russell's lecture, "it's turtles all the way down": it is rhetoric all the way down.[89]

86 Olson, *World on Paper*, 160–179.

87 *Ibid.*, 169. John Knox's famous 1561 exchange with Mary, Queen of Scots, illustrates the early Protestant attitude to which Olson refers. Mary: "Ye interpret the Scriptures in one manner, and [Roman Catholics] in another; whom shall I believe and who shall judge?" Knox: "The Word of God is Plain in Itself" (*History of the Reformation in Scotland*, vol. 2, 18).

88 Olson demonstrates the ways in which early modern experimentalists constructed "objectivity" in *World on Paper*, 171–172.

89 S. Hawking, *A Brief History of Time*, 1. Olson's thesis owes much to the linguistic turn in philosophy and something also to Roland Barthes's hyperbolic formulation of reader-response criticism in "The Death of the Author," 142–149.

Such well-behaved congregations of proper readers, needless to say, were easier to imagine than to create, a problem not lost on the evangelical reformers or new scientists. A priesthood of some believers, not all believers, is the less than ideal compromise reached by most of the Protestants, when it finally came time to empower the masses to interpret Scripture for themselves, and we discover a similar and not entirely unreasonable compromise in the Royal Society's deliberations about the public's role in validating experiments. Of course, these compromises were not always just. Indeed, perhaps they were seldom just, due in large part to the period's cultural blind spots, of which there were many. In a well-documented case, the coruscating discoveries submitted by Dutch commoner Anthonie van Leeuwenhoek to the *Philosophical Transactions*, for example, had first to be vouched for by persons of more noble birth than his own, an instructive episode in the early history of the peer reviewed journal.[90] Nonetheless, the Society—to its credit—did publish Leeuwenhoek's innovative work, to the betterment of humanity.

Satire and the New Experimentalism

The mismatch between the Royal Society's decidedly public idealism and the less-than-ideal situation on the ground unsurprisingly produced a climate for satire. Indeed, wits took aim at the Fellows for a variety of perceived hypocrisies and shortcomings. There was, for example, the Society's seemingly impractical obsession with mechanical novelties—a crucial machine for "the speedy blanching of Hasle-Nuts," as the libertine satirist Ned Ward amusingly highlights,[91] or the infamous parrot-killing "pop gun," the nomenclature on which Hobbes settled in describing the Hooke-Boyle air-pump.[92] More generally, critics accused the Fellows of an impractical obsession with the impractical itself, all the more concerning because of the scientists' brave-new-world rhetoric, one that promised to recuperate what was lost in Eden.[93] Perhaps nowhere do we see a more stinging censure of scientific triviality than in *Gulliver's*

90 D. Lux and H. Cook, "Closed Circles or Open Networks? Communicating Science at a Distance during the Scientific Revolution," 188–189. See also N. da C. Andrade, "The Birth and Early Days of the *Philosophical Transactions*," 9–27.

91 N. Ward, *The Fourth Volume of the Writings of the Author of the London-Spy*, 60.

92 Quoted in Shapin and Schaffer, *Leviathan and the Air-Pump*, 348.

93 L. Walsh, "The Prophetic Ethos of the Early Royal Society," 58–83. For the Society's various appeals to utility, see Hunter, *Science and Society*, 87–112; Gaukroger, *The Emergence of a Scientific Culture*, 36–40; Stewart, *The Rise of Public Science*, xxii, 3–30 (esp. 13). See also Sprat, *The History*, Epistle Dedicatory addressed to Charles II; and the full title of Glanvill's *Plus ultra*, referring to "some of the most Remarkable Late Improvements of *Practical,*

Travels (1726), where the men of Laputa spend their days extracting sunbeams from cucumbers and softening marble for the use in pillows. Exactly how such experiments add up to the betterment of cities we do not know. Shadwell performs a similarly scathing critique in *The Virtuoso* (first staged in 1676), where Sir Nicolas Gimcrack—the quintessential experimentalist—bottles air from various regions of the country, storing it like wine in his cellar, and transfuses the blood of sheep into people, when he is not perfecting his method of swimming on dry land.[94]

Of all the impractical inventions, however, one in particular stands out, probably because it so obviously relates to the history of scholarly communication. This is John Wilkins's universal language scheme, expounded for many years by Wilkins before it was officially published in 1668 for the Royal Society under the title *An Essay Towards a Real Character and a Philosophical Language*. Robert E. Stillman in this volume examines the proto-positivistic rhetorical enterprise, revealing Wilkins's presuppositions about the nature of words and things. In what is undoubtedly one of the great ironies in the history of rhetoric, Wilkins produced a universal language genuinely operable for exactly one person in all of human history, namely Wilkins. There is a lesson in that. One of the earliest and most discerning satires against Wilkins's nascent iterations of the universal scheme comes by way of the anonymous "The Ballad of Gresham College" (1661), which precisely captures the problem of comprehension inherent in the master plan:

> A doctor counted very able
> Designes that all mankind converse shall,
> Spite o' th confusion was at Babell,
> By a character called universall.
> How long that character will bee learning,
> That truely passeth my discerning.[95]

Useful Learning." Glanvill (*ibid.*, 93) specifically highlights Boyle's potent demonstration of the *Usefulness of Natural Philosophy* (1663).

94 Gimcrack functions best in the chthonic world of spiders and weeds, the latter of which he makes "speak eloquently and by a noble kind of prosopopeia instruct mankind," in T. Shadwell, *The Virtuoso*, 18–19. Shadwell also discerningly makes fun of the obscurity produced when Robert Boyle and Robert Hooke strove to write about material matters in plain English. What are Latinisms such as "empyreumatical" (from Boyle's *Sceptical Chymist*, 1661) and "orbiculation" (from Hooke's *Micrographia*, 1667) if not funny? See Vickers, "The Royal Society and English Prose Style," 35–37. See also Gaukroger, *The Emergence of a Scientific Culture*, 37.

95 E. Sherwood Taylor (ed.), "An Early Satirical Poem on the Royal Society."

The language system materialized but was never seriously practiced, and it quickly took its place among the many other oddball relics in the Royal Society's curio cabinet. Moreover, Swift's sages of Laputa might well have rendered the universal language program obsolete, because they advanced one that moved beyond it: "An Expedient was therefore offered, that since Words are only the Names for *Things*, it would be more convenient for men to carry about them such *Things*, as were necessary to express the particular Business they are to discourse on."[96] But the problem here, too, becomes one of impracticality, not to mention absurdity. The packs of things grew heavy, and the conversations—cumbersome: "Sages almost sinking under the weight of their Packs, like Pedlars among us; who, when they met in the Streets, would lay down their Loads, open their Sacks, and hold Conversation for an Hour together."[97]

Shadwell, Ward, and Swift paint vivid pictures of the experimentalist lost in the minutia of material existence, and in the case of Shadwell's *Virtuoso* carnal existence too (there is a thinly veiled subtext in the play impugning Hooke's indiscretions with his housekeeper).[98] The new scientists, they argued, were unable to arrive at big questions or big answers, but all too capable of inventing machines for the speedy blanching of hazelnuts. This line of criticism is perhaps the most serious of all in the satires against the experimentalists, not that they were incompetent per se, but rather that they were unhinged somehow from the important dimensions of life, an always-timely caution for research scientists and academic philosophers in general.[99]

---

96 Swift, *Gulliver's Travels*, Book 3. See also Nate's conclusion on *Gulliver* in this volume.

97 Swift, *Gulliver's Travels*, Book 3.

98 Hooke first heard reports of Shadwell's *Virtuoso* at Garaway's coffeehouse. He attended the play a week later and soon after wrote in his diary, "Damned Doggs. Vindica me Deus. People Almost Pointed" (June 2, 1676).

99 One of the most effective responses to the accusations of minutia-mongering arrived in the form of the Boyle Lectures, which are better described as sermons than lectures. Boyle and other members strictly forbade religious quibbling during meetings of the Society, but the lecture series had as a key purpose the defense of Anglicanism. One of the most successful products from the lecture series was William Derham's *Physico-Theology. Or, A Demonstration of the Being and Attributes of God, from His Works of Creation* (1713), which began as Boyle lectures in 1711 and 1712 and became afterwards one of the most widely sold works of natural philosophy in the eighteenth century (Stewart, *The Rise of Public Science*, 64–65). Regardless of the particular religious agenda behind the Boyle Lectures, a broader edifying point emerges: Boyle saw fit to endow a formal lecture series by which science would be connected to pressing daily concerns in the form of public exposition, and no concern was of greater importance in the early Enlightenment than that of

A final theme in the early satires against the new science requires our attention, and this is the persistent connection made between experimentalism and the traveling circus in its various forms. The Cambridge Platonist Henry More, for example, uses such an analogue when he damns with faint praise the new empiricists' "mechanical kind of Genius that loves to be tumbling off, and trying tricks with the Matter (which they call making Experiments)."[100] Acknowledging scientific ingenuity, More nonetheless sardonically connects the experimentalists to London's stage itinerant, where medicine show funambulists perform wonderful feats of agility for a small fee. Hobbes uses a similar trope derived from the circus, suggesting that Hooke and Boyle "display new machines…in the way that they behave who deal in exotic animals, which are not to be seen without payment."[101] P.T. Barnum, the most infamous retailer of strange sights, would later perfect the rhetoric of dodgy showmanship, but Hobbes disdainfully suggests that the Royal Society was well on its way. An anonymous critic of Desaguliers pursues a comparable thesis, declaring Desaguliers's performances in the 1720s to be nothing but "tricks and talk," interrupted occasionally by "Attraction, Friction, Action, Re-action," and "Suction."[102] Here, the demonstration of new science proves indistinguishable from the demonstration of Three-card Monte.

That satirists found comparisons between the new science's emerging mode of scholarly communication and the traveling carnival's mode of quasi-scholarship is not unexpected, given the circumstances in London. Shapin reminds us that the early Royal Society's experimentalists were itinerant scientists, similar in many respects to itinerant preachers and, of less repute, itinerant apothecaries, who sold their elixers on makeshift stages throughout the city, sometimes accompanied by a piebald assembly of characters, the perennial motley crew of fire spitters, grimace artists, harlequins, monster mongers, jugglers, puppeteers, and fortune tellers.[103] John Gay vividly captures the

religion. Notably, the Boyle lecture series was revived in 2004, not for the purpose of defending the Anglican faith against deists and other infidels (the explicit purpose of the initial lectures), but rather for the purpose of exploring relevant intersections between science and religion in our globalized context. The first volume produced by the lectures is R. Manning and M. Byrne, eds., *Science and Religion in the Twenty-First Century* (2013).

100 H. More, *Enthusiasmus triumphatus*, 50.

101 Quoted in Shapin and Schaffer, *Leviathan and the Air-Pump*, 348.

102 Anon, *Tunbridgialia. Or, the Tunbridge Miscellany for 1730* (London: T.B., 1730), 12. On Desaguliers's role in Richard Steele's wonderfully interdisciplinary Sensorium, see Stewart, *The Rise of Public Science*, 132.

103 S. Shapin, "The House of Experiment in Seventeenth-Century England." On a related note, Rob Iliffe rightly notices that "there were no rigid boundaries between the domains

scenes near which, or through which, patrons walked to and from the new experimentalists' recitals:

> The Mountebank now treads the Stage, and sells
> His Pills, his Balsoms and his Ague-spells;
> Now o'er and o'er the nimble Tumbler springs,
> And on the Rope the ven'trous Maiden swings;
> Jack Pudding in his parti-coloured Jacket
> Tosses the Glove and jokes at every Packet.
> Of Raree-Shows he sang and Punch's Feats;
> Of pockets picked in Crowds, and various Cheats.[104]

The Royal Society performed and advertised new science amongst this clamor, and—importantly—was often seen as a part of it for the first several decades, not only by the satirists but also by the broader educated public for whose attention they were vying. The favorite social outlet of the experimentalists, the coffeehouse, simply reinforced this association of ideas between the rhetoric of science and that of the carnival's "Raree-Show." "Had not my Friend told me 'twas a Coffee-House I should have took it for Quacks-Hall [i.e., Gresham College]," Ward quipped, the "Walls being hung with Gilt Frames, as a Farriers-shop with Horse-shoes; which contained abundance of Rarities" such as "Aegyptian Mummies, Old Musty Skeletons," a "Unicorns Horn," and other "Antiquated Trumpery."[105]

Members of the Society seemed unwilling to cede the fact that no strict border existed in the minds of the public between their experimental enterprises and the enterprises of traveling apothecaries, which perhaps explains Hooke's "shock" at seeing Boyle's austere books "exposed in Moorfields on the railes," presumably next to such classics as the anonymously penned *Bugger Cantus Buggerantissimus* and William Yworth's quack best-selling remedy manual *Chymicus Rationalis. Or, the Fundamental Grounds of the Chymical Art Rationally Stated and Demonstrated* (1692).[106] Among other things, the *Chymicus* provides a

---

of natural philosophy, banausic culture, and construction work on which Hooke was engaged," in "Lying Wonders and Juggling Tricks," 185–211.

104 J. Gay, *The Shepherd's Week in Six Pastorals*, 58.

105 Ward, *Fourth Volume*, 60–61. Ward refers specifically to Don Saltero's coffee shop. For an overview of curio cabinets, wonders, rarities, and the rise of modern science, see L. Daston and K. Park, *Wonders and the Order of Nature*. The King James translation of the Bible gives us a unicorn in Psalm 22:21.

106 Quotations taken from E.L. Furdell, *Publishing and Medicine in Early Modern England*, 119; Ward, *Fourth Volume*, 4.

panacea deemed "Panchymagogon or Catholicon, purging all Humors in the Body of Man," which one ingests at one's own risk, obviously. And neither were the new experimentalists—in their efforts to separate themselves from the London street performers—helped by Swift's manic Tubster, who, in an instructive moment of immense conflation, called out to his many friends at "Will's Coffee-House, and Gresham College, and Warwick-Lane [where the Royal College of Physicians met], and Moor-Fields," the last of which was known for traveling medicine shows and the Bedlam insane asylum.[107]

But such interplay between experimentalism and the back-alley fair should not cause us to assume that the Fellows' most pressing public relations task was to clarify their program contra the London street. We are wise to recall, for example, that Queen Anne had as her closest medical advisor the sham doctor William Read, whose medicine-show balms, pills, and panchymagogues are recorded in his widely circulated book, or advertisement,[108] as the case might be: *Read's true and faithful experiments, lovingly communicated for the good of his countrymen, being a catalogue of those medicaments he sold off his stage during the time of his eighteen years traveling in England, Scotland, Ireland, and many foreign kingdoms.*[109] Even more to the point, Charles II—*Fundator et Patronus*—kept in his service a quack "Chemist from France and patronized a little laboratory under his closet," where Pepys saw "a great many chymical glasses and things, but understood none of them."[110] The irony, or incongruity, did not go unnoticed of a Charles who "mightily laughed" at the Fellows for spending time only in weighing air"[111] while simultaneously frequenting the Dr. Bendos of the world.[112] The Royal Society's attempt to construct and

---

107 Swift, *Tale of a Tub*, 181.

108 As W.F. Bynum noted in relation to early modern medicine show books, "almost any publication—particularly on a practical topic—could be seen as a form of advertising" ("Treating the Wages of Sin," 14).

109 See H. Ormsby-Lennon, *Swift and the Quacks*, 169.

110 C.J.S. Thompson, *Alchemy and Alchemists*, 146.

111 Pepys, *Diary*, Feb. 1, 1664. See also E.S. de Beer, "King Charles, *Fundator et Patronus* (1630–1685)." Charles also reportedly referred to the Fellows as "my ferrets" (see Gaukroger, *The Emergence of a Scientific Culture*, 36), but this is apparently due to a misunderstanding. See M. Jenkinson, *Culture and Politics at the Court of Charles II*, 2, n. 7, with reference to W.E. Knowles Middleton, "What Did Charles II Call the Fellows of the Royal Society?"

112 John Wilmot, Second Earl of Rochester, in a brilliant piece of performance art aimed simultaneously at the credulity of the London street and of the English crown, donned a costume and declared himself in 1675 to be Dr. Bendo. He appeared on London's itinerant stages "in an old overgrown Green Gown which he religiously wore in memory of Rabelais his Master put on at the reception of his Doctor's Degree, at Monpellier, lyned through

empower a modern literate audience, in other words, was not simply directed toward a group of publicans, but rather it was directed toward everyone who had ears to hear.

The Royal Society's Intellectual Milieu

The Society's vision of the modern world, and of modern scholarly communication advanced therein, understandably failed to please everyone. Some were more displeased than others. Against whom did the Royal Society's experimentalists most vehemently contend for legitimacy, epistemic and cosmological, in late seventeenth-century England? Setting aside the hordes of charlatans and crackpots who resisted the Society simply on the grounds that the Fellows were bad for the snake oil business, we are left with an interesting group of alchemists, bourgeoning theosophists, Aristotelians in denial, angelologists, and free thinkers, not to mention the many theologians—of various persuasions—who at times expressed sympathy toward aspects of the Society's experimentalism but nonetheless had questions about where things were heading.[113] In short, the Royal Society's Fellows operated within a crowded public square. Why the experimentalist's version of natural philosophy should win out remained an open question and by no means pointed to a foregone conclusion. Arguments needed to be made, and public demonstrations needed to be successful, which explains why the members fretted so much over the many glitches in Hooke's experimental machinery. Much was at stake. "Fence in public, if you do well, otherwise not"[114]—this is the instruction that William Petty, founding member of the Royal Society, left for his son, and—in essence—it became the Royal Society's informal public relations memorandum to itself. To the broader point: the Fellows were met with serious intellectual opposition in the form of others who claimed better, more productive modes of

with exotick furrs of diverse colours," and about his neck on a "Massy Gold" chain hung "a Magnificent false Medal sett round with glittering Pearl, rubies, and Diamonds of the same Cognation…which the King of Cyprus (you must know) had given him for doing a signal Cure upon his darling Daughter the Princess Aloephangina" (see T. Alcock, *The Famous Pathologist, or The Noble Mountebank*, 29). So convincing was Wilmot's disguise that crowds began to patronize the good Dr. Bendo.

113 There were, of course, other clerics—Anglican and not—who strongly disapproved. Robert South, for example, described the Royal Society's attempt at "new modelling the whole world" as "diabolical," a quotation from a sermon preached at Westminster Abbey in 1667, as cited by Gaukroger, *The Emergence of a Scientific Culture*, 36.

114 W. Petty, *The Collected Works of Sir William Petty*, 302.

experimentalism, and the Society shaped their notion of scholarly communication in juxtaposition to this wide array of opposing worldviews.

Of the Aristotelians much has already been said, perhaps most cogently by Peter Dear in the landmark article appearing in this volume. The new philosophers' general attitude toward the old Aristotelian paradigm (if such a generalization is permissible) finds an especially terse and eloquent expression in "The Ballad of Gresham College":

> Those new colleagues doe assure us,
> Aristotle's an Asse to Epicurus.[115]

Such an attitude is a world away from the conditions at Oxford in 1350, where Bachelors and Masters of Arts who did not follow Aristotle's philosophy were subject to a fine of five shillings for each point of divergence, as well as for infractions of the rules of the *Organon*.[116] But by the middle of the seventeenth century, and after Bacon's revolutionary *New Organon*, the Aristotelian paradigm had fallen into serious decline, Aristotle's non-experimental theory of rhetoric included, and the Society simply accelerated the pace by which Aristotle as such would become increasingly peripheral on matters of rhetoric and science. As a useful example of how the new philosophers went about marginalizing Aristotle, consider in particular Glanvill's *Scepsis scientifica* (1665), one of the earliest and best rejections of the Aristotelian worldview by a member of the Royal Society.[117] Glanvill levels two related charges against the Stagirite's influence: first, that it is ancient and bookish, which puts it on the wrong side of the revolution, and secondly, that such natural philosophy was well-used by the late medieval Catholics, which inherently created suspicion—compounded further by the fact, or problem, that James II waited in the wings.

Of the crepuscular worlds of alchemy and astrology there is too much to say.[118] Maybe it is enough for our purposes to know that these areas of inquiry—and philosophies of scholarly communication contained therein—exerted a profound influence on the culture out of which the Royal Society's experimentalists emerged. Some basic details will illuminate the point. For example,

115 E. Sherwood Taylor (ed.), "An Early Satirical Poem on the Royal Society."

116 K. Devlin, *The End of Logic and the Search for a New Cosmology of Mind*, 21.

117 On the rise of Epicurean-style materialism in seventeenth-century England, see L. Feuer, *The Scientific Intellectual*, 23–82; R. Kroll, *The Material World*.

118 For an introduction to this topic, see—for example—C. Webster, *Magic and the Making of Modern Science*, Vickers (ed.), *Occult and Scientific Mentalities in the Renaissance*; Stark, *Rhetoric, Science, and Magic in Seventeenth-Century England*.

other than the Bible, the astrological almanac was the most widespread form of literature in early modern England.[119] Its only true rival was the sermon. As for alchemical communiques, John Ferguson was one of the early scholars to point out that more alchemical books appeared in England between 1650 and 1680 than in all the time before and after those dates.[120] Add to this detail the fact that many of the landmark Renaissance alchemical books appeared in English during this period (e.g., Agrippa, Paracelcus, della Porta, Fludd), and we arrive at a mid-seventeenth-century English culture saturated with the alchemist's idiom.[121] This is to say nothing of Newton's interest in alchemy. Groups of astrologers and alchemists, too, began to hold meetings and form societies in the neighborhoods of the new experimentalism. Between 1649 and 1658, for example, a group of more than forty astrologers and their acolytes formed a society and met annually in London, "blessed by a sympathetic clergy."[122] William Lily and John Gadbury, the nonconformist and the Royalist astrologer respectively, we should recall as well, commanded substantial audiences in the city and had large client lists.[123] In the 1660s, a society of chemical physicians—"with an essentially gnostic theory of medical knowledge"—was tentatively formed in opposition to the Galenic College of Physicians, another upstart scientific endeavor.[124] Rosicrucian-related activity also saw a striking upsurge, due in part to the revival of Oswald Croll's work and to the publication of many other texts, including John Heydon's nearly impenetrable *Theomagia* (1663), a Rosicrucian guidebook.[125] All of this is to say that the Royal Society's brand of experimentalism developed in the context of a thriving, vibrant milieu in which various systems of experimental inquiry circulated through texts and societies and advanced divergent models of scholarly communication, many of which involved sharp codes of exclusivity, initiation rites, and—in some cases—the sporting of amulets.

---

119 B. Capp reports that one of every three English households in the middle of the seventeenth century owned an almanac, in *Astrology and the Popular Press. English Almanacs 1500–1800*, 22–23.

120 J. Ferguson, "Some English Alchemical Books," 5.

121 Cornelius Agrippa's *Occult Philosophy* in 1651; Paracelcus's *Works* 1657; Giambattista della Porta's *Natural Magic* in 1658; Robert Fludd's *Mosaicall Philosophy* in 1659.

122 Vickers, "The Royal Society and English Prose Style," 53.

123 Samuel Butler's character of Sidrophel in the hugely popular *Hudibras* (1684) is based loosely on Lily and probably did more to discredit academic astrology in the Restoration than all of the sober arguments and sermons combined.

124 A. Coudert, *The Impact of the Kabbalah in the Seventeenth Century*, 153.

125 Croll's "A treatise of Oswaldus Crollius of signatures of internal things" was reissued in *Royal and Practical Chymistry* (London, 1670).

As a brief *caveat lector*, we note also that most of the period's astrologers would recoil at the thought of being lumped into the same group as the natural magicians. This objection is a fair one. We place the astrologers with the alchemists not because they fit but because there is no other place to put them. C.S. Lewis shrewdly noted that the primary difference between the Renaissance natural magician and astrologer is that the former views the human being as having immense power and cosmic potential, while the latter, the astrologer, is all too happy to argue for human impotence.[126]

The theosophists, or spiritual experimentalists, were all too happy to distance themselves from the Hooke-Boyle brand of empiricism. We discover a particularly instructive example in *The Philadelphian Society of London*, shepherded by Jane Lead, probably the most important female religious figure in late seventeenth-century England, with the possible exception of Margaret Fell.[127] Notably, the Philadelphians published a scholarly journal entitled *Theosophical Transactions* (1697), designed specifically as a counterstatement to the Royal Society's *Philosophical Transactions*. Among other things, the theosophists criticized the Society's Fellows for their overconfidence in "Creaturely Evidence" and "the dark Mists of [rational] sense and worldly wisdom."[128] The Philadelphians gathered to perform—and report upon—experiments of a different sort entirely: the revelation via contact with spirits, the séance, the vision quest. They drew much of their inspiration from the pietist Jacob Boehme, whose complete works were Englished between 1644–62, and the occult philosopher John Pordage, whose *Theologia mystica* (1683) created a sensation in the circles of esoterica and caused less than a ripple amongst the modern physico-theologians. Understandably, predictably, Lead and her devotees were met with a sharply mixed reception, as, incidentally, were the Quakers and other related groups who embraced gender equality in the pulpit and inner-light equality in the heart. John Locke deemed the Philadelphians to be "not comprehensible," adding that purpose of the new science was to differentiate "between the enlightened and dark Parts of Things," the latter of which belonged to the Philadelphians, and they to it.[129] William Law, Emanuel Swedenborg, and William Blake believed differently, crediting Lead as a visionary in the realm of cosmology. There is, for instance, something of the Philadelphians in Blake's warning against the traps of reason: "He who sees the infinite in all things sees

126 Lewis, *English Literature of the Sixteenth Century, Excluding Drama*, 6.

127 See P. McDowell, "Enlightenment Enthusiasms and the Spectacular Failure of the Philadelphian Society."

128 Lead, *Revelation of Revelations*, 31; Lead, *The Heavenly Cloud Now Breaking*, 11.

129 Locke, *Essay Concerning Human Understanding*, 1.1.7.

God. He who sees the Ratio only sees himself only."[130] And it is probably correct to say that Werner Heisenberg's temperament was as much Philadelphian as it was newly experimental when he declared, "What we observe is not nature herself, but nature exposed to our method of questioning."

But more to the point at hand: if we open any of Lead's seventeen books (most of which also appeared in German and Dutch translations), we quickly discover a prose style similar to John of Patmos's in the Book of Revelation ("Revelation"—"a book in which Saint John the Divine concealed all that he knew," as Ambrose Bierce amusingly remarks in *The Devil's Dictionary*).[131] Lead writes in the impetuous language of the mystic, which is not the style of new experimentalism. The Oxford-educated nonjuror and medical doctor Francis Lee defended Lead's rhetoric by noting that the "stile and manner" of her writings were simply "not suitable to the Genius of this Polite Age," but he added a warning: it is foolish to reject Lead's parabolic, emblematical style "as if one should condemn Euclid for writing Jargon, and for being the Author of a company of extravagant Whimsies in the Mathematicks; without having ever Thoroughly Read, or Comprehended the very first Definitions and Postulates, which are the foundation to the whole Art."[132] In Lee's defense of the Philadelphians' rhetoric we catch a glimpse of nearly every criticism leveled against the Royal Society's new mode of scholarly communication by disenfranchised groups, and this is the basic observation that idioms and methods necessarily vary. One person's jargon is another person's treasure, a perfectly salient point, of course, but one that also potentially leads to episodes such as the Sokal hoax.[133] Nonetheless, Lee's apology proves illuminating because it shows a conflict between two sincere modes of scholarly communication in the early Enlightenment. Lee judges one to be more suitable for the Augustan Age, which is to say that Lee understood all too well that the Philadelphians were losing the public relations argument, while the Royal Society's mode of scholarship was waxing.

Angelology once again entered the public imagination, if it ever exited, when Meric Casaubon in 1659 published John Dee's *A True and Faithful Relation of What Passed for Many Years between Dr. John Dee and Some Spirits*, informally known as the angel conversations. The book was an instant success: theosophists, shamans, devotees of Jacob Boehme, mediums, ghost hunters, and pietists of various sorts added another philological arrow to their quiver. But

130 Blake, *The Complete Poetry and Prose of William Blake*, 2.

131 Bierce, *The Devil's Dictionary*, 292.

132 Lee, "Introduction," *Heavenly Cloud*, A1v.

133 A. Sokal, "Transgressing the Boundaries."

was Dee's work in any sense "natural philosophy?" Deborah Harkness argues the affirmative, suggesting that Dee, after encountering several obstacles in his alchemical-scientific inquiries, looked for a new empirical method of gaining insight into the natural world.[134] Thus, he began systematically to experiment with the scrying stone, if we are to believe Dee. Accordingly, eventually, he contacted and communicated with angels, not for the purpose of finding buried treasure (that would come later), but rather for the purpose of gaining access to knowledge beyond the present human reach. This is not an unusual plan in the history of thought—to seek inaccessible knowledge by doing something unconventional.[135] In this context of pursuing nature's secrets in the early modern world, Dee's scrying stone might therefore temporarily be placed alongside the telescope and microscope. All three, if wielded properly, promised a vision of reality beyond the normal senses, and all three generated immense interest in mid-seventeenth-century England. Perhaps this is where the similarity ends, however. The microscope and telescope utterly revolutionized modern science; Dee's scrying stone is on display in the British Museum's cabinet of curiosities.

Dee's mode of experimentalism, and of scholarly communication therein, thus did not develop into a branch of modern science. The angel experiments were impossible to replicate, which was an insurmountable problem, though much testimony from Dee and others proved that something peculiar happened. Yet testimony by itself, as the early moderns knew well, was subject to the predictable momentum of credulity, already documented in the period's literature on enthusiasm.[136] What—exactly—happened in the spiritual conversations, Dee's theater, remained in the minds of new philosophers a matter of faith and conjecture, a Rorschach test of a sort that found no solution in Dee's rigorously descriptive field notes on his interactions with the angels who might have been angels, devils, or chimeras of the imagination. These field notes simply did not pass the rhetoric-of-presence test put to them by the Fellows. Indeed, angel conversations, whether Dee's or a theosophist's, no matter how vividly portrayed, became one more foil against which the experimentalists defined their project.

The last counter-public under consideration, the free thinkers, are—like the alchemists—difficult to categorize: there is a profound diversity of thought

134 D. Harkness, *John Dee's Conversations with Angels*, 98–130.

135 Saul, for example, contacted the Witch of Endor (Samuel 28: 3–25).

136 M. Casaubon, *A Treatise Concerning Enthusiasm* (1654). Hume's rejection of testimony in "Of miracles" had not yet occurred to the members of the Society, but *nullis in verba* is a step in Hume's direction, if not a leap.

within the grouping. Swift defined "free thinkers" as those with a lunatic propensity to speak immediately whatever came to mind, but here, for our purposes, we denote those philosophers with a vested interest in defending skepticism, deism, and, worst of all by the Royal Society's standards, atheism. Indeed, part of the Society's foundational purpose was to investigate nature in order better to understand God's handiwork, which therefore allowed an even more worshipful attitude toward life. Nearly every member of the Society expressed what today would be categorized as some form of Christian intelligent design theory. The *Philosophical Transactions* was not prepared to publish an article declaring that God had died or even, for that matter, that God was absent. Of course, some of the Royal Society's Fellows dabbled in free thinking, just as some of the Society's Fellows dabbled in, and dipped in, alchemy, but the overwhelming mainstream of the Society resisted the free thinkers on every important front. As a typical illustration, consider the riddled topic of spiritual causation, an issue that haunted most Fellows well into the nineteenth century. That the Society worried about what they deemed sadducism, the denying of spiritual causation, the existence of angels, and so forth, goes without saying; the early literature is packed full of admonitions against it. Why it was worrisome is perhaps equally obvious: the final consequence of denying spiritual causes, if the syllogism plays itself out, is mordant materialism. Glanvill makes such an argument throughout *Saducismus triumphatus* (1681), where he chides "those Hobbesians" who "slight Religion and the Scriptures, because there is such express mention of Spirits and Angels in them, things that their dull Souls are so inclinable to conceive to be impossible," adding moments later that he sees "a special piece of Providence that there are ever and anon such fresh examples of Apparitions and Witchcrafts as may rub up and awaken their benummed and lethargick Mindes into a suspicion at least, if not assurance that there are other intelligent Beings besides those that are clad in heavy Earth or Clay."[137] Setting the *ad hominem* aside, or maybe not, nearly every member of the Royal Society agreed with Glanvill on this point. Spiritual causes were taken as a given; the real question for the experimentalists involved what spiritual causation meant in terms of the scientific enterprise; to this question there were many answers, William Derham's *Physico-Theology* (1713) among the most popular. But by denying in a seemingly *a priori* fashion preternatural and supernatural roles in material daily life, the free thinkers—by the Fellows' precepts—preemptively disqualified themselves from the new experimentalism. Instead, the free thinkers became synonymous with a kind of disgruntled skepticism, a morality tale in the context of mainstream science,

137 Glanvill, *Saducismus triumphatus*, 16.

illustrating how unbelief toward divine and demonic interventions easily slips into cynicism and then atheism.

What do the free thinkers tell us about the Royal Society's mode of scholarly communication? Perhaps more than anything, the Society's response to the free thinkers proves revealing. We discover that the Society operated within a rigorous set of religious presuppositions and metaphysical parameters. Ergo, the free thinkers indirectly helped to contextualize and historicize the Society's particular brand of experimentalism, in part because they questioned the necessity of some of the theological assumptions underlying the new science, which is to say that they exposed the assumptions to the public, and in part because they offered a nearly endless supply of satirical quips against the Fellows, and more quips still if we count the young Swift as a free thinker, which he might well have been. In short, the free thinkers—like all skeptics—were good at raising their eyebrows.

But authentic places of experiment are always touched by the ridiculous, then and now. Why should the Royal Society's places be any different? Until the very end of the seventeenth century, the Fellows debated the benefits of drinking cow urine.[138] And neither should we lose sight of Newton's obsession with the occult and Boyle's belief that mummified fingers were a useful ingredient in various remedies. As many scholars have shown, early modern experimentalism was by no means a coherent practice, or settled set of methods, well into the eighteenth century. Trial and error requires a lot of error, and there is a reason why the term "habnab" was a synonym for empiricism in the early Enlightenment.[139] But perhaps this much in fairness can be said: the Fellows made many mistakes—some of them spectacularly carnivalesque—in a productive direction, as they invented a way of bringing science to a larger public sphere. And, quite literally, they did *bring* science to the public. The Society took its show on the road and competed against other shows, but with one crucial difference again worth stressing; rather than hocus pocus theatrics wherein the mechanisms of astonishment were well-hidden behind the curtain or underneath the hat out of which the rabbit jumped, the Fellows performed their demonstrations out in the open for everyone to see and replicate, and by "everyone"—as has been shown—the Society meant a small group of

---

138 E. Dolnick, *The Clockwork Universe*, 8.

139 Peter Cole, the Puritan printer of medical advice books and antagonist to the new experimentalists, defined "an Emperick" as one who "gives Physick (Hab Nab as we use to say) releying only on Experience, and what he has seen done before him, not being able to give any reason touching the Disease, its Cause, or Cure." See Cole, "The Printer to the Reader," Ar.

patrons and distinguished guests. In practical terms, however, there was nothing else to be done. The Society could not build a stadium large enough for its critics and competitors, but what they could do—and did with much success—was create a philosophy of rhetoric that delivered something even better (in many respects) than actual witnessing, given that the page preserved the opportunity for virtual witnessing at the audience's leisure. This audience might exist at some indeterminate point in the future or some determinate point across an ocean, or across a language, as Felicity Henderson has recently shown, provided that proper care was taken with the translations.[140] Those who define the terms win the debate, the old adage goes, and the Royal Society's genius was not in creating makeshift experiments, at least not in the beginning, but rather in defining the terms. They fashioned a rhetorical framework, a philosophy of modern scholarly communication by which experiments might be discussed and—of equal import—verified by empowered audiences who also wielded the hermeneutics of objectivity, or what the Society informally deemed the reading of things plainly, whether those "things"—or facts—operated in the Bible or the Book of Nature. Therein we discover the birth of modern rhetoric and the concomitant disinterested readers and writers who participated in it. The Fellows successfully engineered a new species of communication.

A New Enlightenment?

In our own digital age, the Royal Society has continued to develop new methods of communication both within the scholarly community and toward the wider public. While Fellows from the UK and around the world can now log in to the eFellows Room, anyone with access to the Internet can scrutinize the Society's daily communications via Facebook, Twitter, Youtube, and blogs such as "In verba" (posted by the Royal Society's own Science Policy Center). A new open access journal, *Royal Society Open Science*, marks the 350th anniversary of the launch of *Philosophical Transactions* in 2015.[141] Maintaining a high media profile and making good use of new media, the UK academy of science has thus recently renewed its laudable ambition to strengthen the bond between science and society.[142] When the Royal Society celebrated its 350th Anniversary

140 F. Henderson, "Translation Theory and Practice at the Early Royal Society."

141 http://royalsocietypublishing.org/royal-society-open-science.

142 http://blogs.royalsociety.org/in-verba/. By comparison, the international website of the French Académie des sciences appears less colorful and inviting, see http://www.academie-sciences.fr/index_gb.htm.

in 2010, the Society's then president Sir Martin Rees announced that "There can be no better aim, for the next fifty years, than to sustain the curiosity and enthusiasm of our founders, while also achieving the same broad engagement with society and public affairs as they did."[143] Since then, Martin Rees's successor, Sir Paul Nurse, has called for "a culture shift" in order "to reawaken the spirit of the Enlightenment" and the Industrial Revolution and let experimental science help Britain regain its former position as a world-leading "economic powerhouse."[144] Nurse first made this appeal in a lecture broadcasted on BBC One on 28 February 2012, which was immediately made available on Youtube as well as the Royal Society's website. With explicit reference to Francis Bacon, Robert Hooke and the achievements of the early Royal Society, Nurse voiced a strong argument in favor of "reproducible observation and experiment," claiming that because "[science] has always been a useful art," there is no doubt that it can now "help solve the world's biggest problems."[145] In the twenty-first century, arguments based on utility and applicability have increasingly been used to legitimate why research institutions and universities deserve public support and funding. In its yearly announcements of research support, the European Commission asks routinely: "What's the benefit for citizens?"[146] While the arts and sciences taken together have been losing ground to vocational studies both in the UK and worldwide at least since the 1980s, the Royal Society's "Fellowship of the world's most eminent scientists" holds the promise of offering solutions to big-scale problems in health, environmental sustainability and the economy, while at the same time aspiring towards public engagement.[147]

143 See Martin Rees's "Conclusion," in *Seeing Further*, ed. B. Bryson, 469.

144 "The Richard Dimbleby Lecture 2012. The New Enlightenment," 15, 3.

145 Quoted from "The Richard Dimbleby Lecture," 5, 8 and 3. See also *ibid.*, 7, 9 and 12.

146 European Commission's Seventh Framework Programme for Research and Technological Development. The changing research environment has been analyzed in terms of (a) the "steering" of research priorities, (b) the commercialization of research, and (c) the accountability of science, by H. Nowotny et al., "Introduction. 'Mode 2' revisited," 181–186. For a discussion of "usefulness" as "the main qualifying criterion for public funding of education," see H. Small, *The Value of the Humanities* 59–88 (61).

147 Quoted from http://royalsociety.org (accessed 26 June 2013). High promises are given e.g. at http://royalsociety.org/policy/ (accessed 26 June 2013). For the fate of the liberal arts, see A. Kernan's Introduction to *What's Happened to the Humanities*, 6, excerpted in W. Smith and T. Bender (eds.), *American Higher Education Transformed*, 47–48 (48). See also Smith and Bender's Introduction, *ibid.*, esp. 9. M.C. Nussbaum has argued in *Not for Profit*, 2, that "what we might call the humanistic aspects of science and social science—the imaginative, creative aspect, and the aspect of rigorous critical thought—are...losing ground as nations prefer to pursue short-term profit by the cultivation of the useful and

The strategic plan for 2012–2017 declares that "The Society is committed to ensuring that everyone has the opportunity to appreciate the value of and engage with science, whether through top quality formal education or through other resources and events."[148]

The current public relations strategy can be traced back to a 1985 report, which urged that "the Royal Society should make improving public understanding of science one of its major activities."[149] Defining science broadly to include both "the systematic investigation of the natural world and the practical application of knowledge derived from such investigation," the so-called Bodmer Report specifically addressed how to win the support of industry owners as well as policy-makers and their voters.[150] As part of this goal, individual scientists were encouraged to "learn how to explain science simply, without jargon," a goal which revived the ambition of the Society's original founders and reinvigorated the early modern conception of rhetoric as a mode of communication between the learned and the layman.[151] The Bodmer Report was clearly not much concerned with the style of peer-reviewed writing; its concern was rather with an aspect of the rhetoric of science that has to do with fundraising and securing goodwill. In light of the recent economic crises, such efforts have become even more crucial. In their handbook on *The Craft of Scientific Communication* (2010), Joseph D. Harmon and Alan G. Gross maintain that the most important reason why scientists today should seek to popularize their insights is that "such writing leaves a favorable impression on the public—in particular, the citizens whose taxes support much of scientific research. Writing for the general public celebrates science."[152]

With the advent of new digital media, there is even some possibility for interaction and two-way-communication, although in reality feedback is often handled by professional public relations practitioners and technocrats rather than the scientists themselves.[153] As the former CEO of the British Science Association Sir Roland Jackson has pointed out, "public dialogue, reflecting an

highly applied skills suited to profit-making." The alleged "consumerist [ethos] of increased prosperity" is also critiqued by S. Collini, *What Are Universities For?* esp. 90.

148 "Strategic Plan 2012–2017," at http://royalsociety.org/about-us/governance/.

149 The Royal Society, "The Public Understanding of Science," 6. The background for this report is analyzed in N.J. Russell, *Communicating Science*, 70–74.

150 Bodmer was the name of the chairman of the working group responsible for the report, Dr W.F. Bodmer. See the Preface to "The Public Understanding of Science," 5. Quoted from *ibid.*, 7.

151 *Ibid.*, 6; Howell, *Logic and Rhetoric in England*, 3–4.

152 Harmon and Gross, *The Craft of Scientific Communication*, 135.

153 See the conclusion to M. Pieczka and O. Escobar, "Dialogue and Science," 123.

actual public involvement in thinking about priorities and directions for science and technology" has yet to be fully developed.[154] The preferred mode of science communication is still the traditional "educate and inspire" mode, as seen in the numerous festivals and fairs such as the British Science Festival, the Big Bang Fair, and the Royal Society's annual Summer Science Exhibition.[155] Whereas one may compare these events to the public lectures and demonstrations taking place in the long eighteenth century, today's combination of video and information technology ensures that there are more ways to enjoy the "fascinating" and "fun" performances of science.[156]

The new technologies also inspire truly innovative methods for the speedy dissemination of research findings and methods. The peer-reviewed *Journal of Visualized Experiments* (JoVE), for example, publishes video-protocols from an international community participating in biological, medical, chemical and physical research.[157] JoVE offers an interesting contemporary counterpart to the early Royal Society's struggle to find an alternative to attendance at the original locus of scientific *factum* or fact-making. Indeed, such "show-and-tell" science exceeds by far the seventeenth-century experimentalists' ambition to facilitate not only the understanding but also the efficient reproduction of experiments and techniques.[158] Additionally featuring videos for use in science education, the JoVE website, like the Royal Society homepage, offer newsfeeds with recent research results.

In today's visual and electronic culture, the rhetoric of science increasingly takes shape through ongoing virtual performances, on flashing webpages, in webcasts of public lectures, and in TV shows hosted by such celebrated popularizers as the particle physicist Professor Brian Cox.[159] In medicine, the use of artistic, colorful and often animated pictures of microscoped cells and cancers seems designed not only to explain how the body works, but also to convince the audience of the wonders of science.[160] Such moving images catch the audi-

154 R. Jackson, "Ten Years of Public Engagement."

155 According to Jackson, *ibid.*, there was a "massive growth in science festivals" in the period from 2002–2012.

156 "Strategic Plan 2012–2017."

157 http://www.jove.com/.

158 Quotation taken from A. Bonasio, "Science Show-and-Tell," Mendeley Blog, 15 March 2013. Bonasio reflects on the foundation of JoVE in 2006, but without any reference to early modern science.

159 Cox is also a Royal Society University Research Fellow, see http://royalsociety.org/events/2013/brian-cox (accessed 26 June 2013).

160 The effects of the use of new imaging technologies in research, medical practice and science communication have been debated in a short documentary by Anwar Saab,

ence's attention like nothing else, but the question remains whether the spectators have the competence to evaluate the findings thus presented. One may ask as well whether the increasingly sophisticated presentation methods in scholarly communication ought to be subjected to the same level of critique as other forms of infotainment. From the scientists' perspective, there is also the question of how to balance the need to make science appear "exciting and interesting" against the equally pressing concern with developing "deeper relationships" with citizens, for the sake of both science and democracy.[161] The Baconian ideal of openness has definitely met with some challenges in a culture which—as the Royal Society's report on "Science as an open enterprise" maintains—will increasingly "allow citizens to examine scientific information at the click of a mouse."[162] Insisting that feedback is welcomed, the scientific community is grappling with how to handle "ridicule" and "belittlement".[163] Speaking of social media such as blogs and Twitter, Roland Jackson has called for "mutual respect" between "practising and communicating scientists" and those commenting on the nature of science, whether they be philosophers, historians or social scientists.[164] When it comes to the mass media, Paul Nurse has voiced a concern also heard among the early Royal Society, that scientific debate could be "distorted" by "ideology, politics, and religion."[165] A Nobel-Prize winning geneticist who has received hate-mails from people objecting to stem cell research, Nurse has found reason to insist that decisions should be based on "scientific facts and analysis" rather than polemical and "preconceived" opinions.[166] As Nurse polemicizes: "The time for politics is after the science not before."[167]

The Good, the True and the Beautiful. A Film about Scientific Imaging (2012), based on the research project "Inside Out. New Images and Imaginations of the Body" at the Norwegian University of Science and Technology. See http://insideoutimaging.wordpress.com/film-project/, and watch the teaser at http://www.youtube.com/watch?v=SR7FhzIL5Bk&feature=youtu.be.

161 Quotations taken from an interview with Paul Nurse, http://royalsociety.org/people/paul-nurse/interview-jan-2011/ (accessed 20 June 2013); and from Jackson, "Ten Years of Public Engagement."

162 http://royalsociety.org/policy/projects/science-public-enterprise/.

163 Jackson, "Ten Years of Public Engagement."

164 *Ibid.*

165 "The Richard Dimbleby Lecture," 17, 15, 9. Two surveys of scientists' views of the media in the UK and US have shown that whereas "scientists are critical of media coverage generally, ...they also tend to rate favorably their own experience dealing with journalists, believing that such interactions are important both for promoting science literacy and for career advancement." See J.C. Besley and M. Nisbet "How Scientists View the Public, the Media and the Political Process," 6–8 (quotation taken from the Abstract on p. 1).

166 "The Richard Dimbleby Lecture," 10–11 (10).

167 *Ibid.*, 9.

Nevertheless, the making of science requires a continual rhetorical effort. As Bruno Latour has argued, "The problem of the builder of 'fact' is the same as that of the builder of 'objects:' how to convince others, how to control their behaviour, how to gather sufficient resources in one place, how to have the claim or the object spread out in time and space."[168] Schaffer and Shapin raise a similar argument on behalf of the seventeenth century polity of science in *Leviathan and the Air-Pump*: "Scientific activity, the scientist's role, and the scientific community have always been dependent: they exist, are valued, and supported insofar as the state or its various agencies see point in them."[169] Taken together, the articles collected in this volume give valuable background on some of these highly contested issues in the history of scholarly communication, issues that are still being debated, such as the question of how to accommodate subjective responses from actual or virtual witnesses, while maintaining science's claim to objectivity.

This Volume

As indicated above, the collection of material in this volume is divided into two parts, one which discusses the Royal Society's rhetoric and views on language from a theoretical perspective and the other presenting case studies of individual founder members' contributions to rhetorical and linguistic theory. Part one begins with an essay placing the Royal Society within the larger perspective of the Scientific Revolution: Peter Dear's "*Totius in verba.* Rhetoric and Authority in the Early Royal Society" (from *Isis* 76,2, 1985). Emphasizing the cooperative nature of natural philosophy, this foundational study highlights the new sense of authority that followed from experiment and observation. Instead of making generalized statements based on ancient authority, the Fellows of the Royal Society typically reported on specific experiences and observations, often by the use of an active voice, as in a first-person narrative. Dear thus treats the matter of literary form or style not in terms of any opposition between science and rhetoric or science and literature, but as an integral aspect of the new ideal of natural inquiry shared among practitioners of the new science in England and Europe, an ideal that was based on experimentation and observation.

In the next chapter, "Rhetoric in the Early Royal Society," Richard Nate argues convincingly that the natural philosophers not only made use of rhetorical

168 Latour, *Science in Action*, 131.

169 Schaffer and Shapin, *Leviathan and the Air-Pump*, 339.

devices in their writings, but they also developed a rhetorical theory of their own (from *Rhetorica movet. Studies in Historical and Modern Rhetoric in Honour of Heinrich F. Plett*, eds. Thomas O. Sloane and Peter L. Oesterreich, Brill, 1999). The overriding principle was that of *aptum* or appropriateness: the ancient ideals of linguistic simplicity and transparency (*perspicuitas*), which were traditionally associated with the middle style, were considered as most suited to scientific discourse. Because the norms of such discourse had yet to be established, texts written in the plain style were often invested with a meta-communicative purpose: by stating his intention of cultivating perspicuity and avoiding obscurity, the author seeks to persuade the reader that his text deserves to be counted as "scientific." The extent to which the Royal Society downplayed the value of the rhetorical discipline, Nate argues, was a consequence of the so-called Ramist reform of education in the late sixteenth century, whereby rhetoric was reduced from a system for developing thoughts and ideas as well as verbal expressions (from *inventio* and *dispositio* to *elocutio* and *actio*) to a system devoted only to the study of style and delivery.

The third chapter in Part 1 is entitled "Language Reform in the Late Seventeenth Century" (excerpted from Ryan J. Stark's monograph, *Rhetoric, Science and Magic in Seventeenth-Century England*, CUA Press, 2009). Stark argues that modern scholarly communication began at that moment when the new scientists disenchanted language. He demonstrates, in particular, how the Royal Society's experimentalists advanced the idea of rhetorical plainness in order to reject natural magic and mysticism. In this context, "plainness" denoted a lack of enchantment in discourse, not an absence of figuration, and it functioned as a shorthand way to dissociate one's eloquence from numinous modes of writing, where tropes had charm, either naturally or preternaturally imbued. Stark suggests that the plain language reforms in late seventeenth-century England ought to be understood as part of this much broader struggle between occult philosophy and mysticism, on the one hand, and, on the other hand, experimental science.

The fourth and final chapter of Part 1 is Alan G. Gross, Joseph E. Harmon and Michael S. Reidy's, "Argument and 17th-Century Science. A Rhetorical Analysis with Sociological Implications" (originally published in *Social Studies of Science* 30,3, 2000). Gross, Harmon, and Reidy examine argumentative techniques in three of the earliest examples of scholarly communication journals: the Royal Society's *Philosophical Transactions*, the Académie Royale des Sciences' *Mémoires* and *Journal des Sçavans*. This comparative and partly quantitative study intriguingly demonstrates that the English experimentalists were more inclined to hedge their conclusions and arguments about facts, often avoiding the tone of epistemological certainty, while the rationalists in Paris were more likely not to hedge conclusions, presenting more often a disposition of certitude. We discover that the presentation of facts in the earliest examples of modern scholarly communication is deeply rhetorical in nature and influenced by the temperaments of the institutions and research programs involved.

Part Two presents four case studies. In the first, "Invitation and Engagement. Ideology and Wilkins's Philosophical Language" (from *Configurations* 3,1, 1995), Robert E. Stillman places John Wilkins's universal language project in the context of Restoration theology and politics, showing that Wilkins aimed to codify the rhetorical ideals and habits of English gentleman scientists interested in increasing commerce, public support for science, and cooperation among the world's experimentalists. The universal language program functioned not as practice but rather as idealistic propaganda, though Stillman painstakingly demonstrates that the universal characters contrived by Wilkins were intelligible and, ironically, modeled on an Aristotelian taxonomical design that recalled the great chain of being, hardly the new world order. That any serious philosopher would have the patience to wield the cumbersome universal characters—Stillman suggests—was unlikely almost from the start.

Next is Frédérique Aït-Touati's sharp-eyed analysis of Hooke's poetics: "'The Spirit of Invention.' Hooke's Poetics for a New Science in *An Attempt to Prove the Motion of the Earth by Observation* (from *Etudes Epistémè* 14, 2008). Aït-Touati shows how Hooke created an innovative mode of writing, a "material poetics of the new science," in order to promote the Royal Society's particular brand of experimentalism. Inspired by Baconian pragmatism, Hooke, she demonstrates, retooled the classical canons of invention, arrangement, and delivery for the singular purpose of effectively discovering, organizing, and presenting the new experimentalist's matters of fact, the seeds separated from the chaff, to use a biblical metaphor from Hooke's preface to *An Attempt to Prove the Motion of the Earth by Observation*, the main text under consideration. In other words, the literary technology of writing becomes an extension of Hooke's instrumental technologies (for example, the microscope, the air-pump), and rhetoric—another utensil by which the experimentalists demonstrated their facts.

Third is Michael Wintroub's "The Looking Glass of Facts. Collecting, Rhetoric, and Citing the Self in the Experimental Natural Philosophy of Robert Boyle" (from *History of Science* 35, 1997). Wintroub shows how *enargeia*, vivid and lively description, functions as the central rhetorical technique in Boyle's experimental natural philosophy and—perhaps by extension—the new experimentalist's rhetoric of fact writ large. Boyle aimed to show, and not tell, audiences about scientific discoveries, using the machinery of writing to create veritable virtual witnesses. For Wintroub, Boyle fashions a provocative rhetoric of presence, attempting to make extant in the most literal sense possible the facts discovered by experimental methods. Thus, Boyle envisions and enacts a mode of scientific writing that functions not as commentary or testimony but rather as demonstration.

In the fourth and final case study, "Science versus Rhetoric? Thomas Sprat's *History of the Royal Society* Reconsidered" (from *Rhetorica. A Journal of the History of Rhetoric* 29, 1, 2011), Tina Skouen reassesses Sprat's *History of the Royal Society* in light of Sprat's handling of the rhetorical tradition. Whereas previous discussions have mostly centered on whether or not Sprat rejects the ancient art of rhetoric, the present study investigates his manner of approaching past authorities. As a writer, Sprat demonstrates the same kind of utilitarian attitude towards the handed-down material in his field of knowledge as he says is characteristic of the Royal Society's natural philosophers. Making good use of Ciceronian ideas, Sprat emerges not as a condemner but as a rescuer of rhetoric. The last article of the volume also establishes a link between the early modern age and contemporary culture by evaluating the Royal Society's 350th Anniversary Campaign of 2010.

References

Agrippa, Heinrich Cornelius, *Three Books of Occult Philosophy*, trans. John French (London: Printed by R.W. for Gregory Moule, 1651).

Aït-Touati, Frédérique, *Fictions of the Cosmos. Science and Literature in the Seventeenth Century*, trans. Susan Emanuel (Chicago: University of Chicago Press, 2011).

Alcock, Thomas, *The Famous Pathologist, or the Noble Mountebank*, ed. Vivian de Sola Pinto (Nottingham: Sisson & Parker for the University of Nottingham, 1961).

Andrade, N. da C., "The Birth and Early Days of the *Philosophical Transactions*," *Notes and Records of the Royal Society of London* 20 (1965), 9–27.

Anon, *Tunbridgialia. Or, the Tunbridge Miscellany for 1730* (London: T.B., 1730).

Barthes, Roland, "The Death of the Author," in *Image-Music-Text*, trans. Stephen Heath (New York: Hill and Wang, 1977).

Battigelli, Anna, *Margaret Cavendish and the Exiles of the Mind* (Lexington: University Press of Kentucky, 1998).

Bauman, Richard, and Charles S. Briggs, *Voices of Modernity. Language Ideologies and the Politics of Inequality* (Cambridge: Cambridge University Press, 2003).

Bazerman, Charles, "The Writing of Social Organization and the Literate Situating of Cognition. Extending Goody's Social Implications of Writing," in *Technology, Literacy and the Evolution of Society. Implications of the Work of Jack Goody*, eds. Michael Cole and David R. Olson (Mahwah: Erlbaum, 2006), 215–239.

——, *Shaping Written Knowledge. The Genre and Activity of the Experimental Article in Science* (Madison: University of Wisconsin Press, 1988).

Bender, Thomas, and Wilson Smith (eds.), *American Higher Education Transformed, 1940–2005. Documenting the National Discourse* (Baltimore: Johns Hopkins University Press, 2008).

Besley, John C., and Matthew Nisbet, "How Scientists View the Public, the Media and the Political Process," *Public Understanding of Science* (2011), published online 30 August 2011, http://pus.sagepub.com/content/early/2011/08/26/0963662511418743.

Bierce, Ambrose, "The Devil's Dictionary (1911)" in *The Collected Works of Ambrose Bierce*, Vol. 7 (New York: Neale, 1911).

Blake, William, *The Complete Poetry and Prose of William Blake*, ed. David Erdman (Berkeley: University of California Press, revised edition 1982).

Bonasio, Alice, "Science Show-and-Tell," Mendeley Blog, 15 March 2013, http://blog.mendeley.com/research-miscellanea/science-show-and-tell/ (accessed 26 June 2013).

Bryson, Bill, ed., *Seeing Further. The Story of Science & The Royal Society* (London: Harper Press, 2010).

Butler, Samuel, *Hudibras* (London: Printed by T.N. for Henry Herringman, 1684).

Bynum, W.F., "Treating the Wages of Sin," in *Medical Fringe and Medical Orthodoxy, 1750–1850*, eds. W.F. Bynum and Roy Porter (London: Croom Helm, 1987), 5–28.

Cantor, Geoffrey, "The Rhetoric of Experiment," in *The Uses of Experiment. Studies in the Natural Sciences*, eds. David Gooding, Trevor Pinch, and Simon Schaffer (Cambridge: Cambridge University Press, 1989), 159–180.

Capp, Bernard, *Astrology and the Popular Press. English Almanacs 1500–1800* (London: Faber, 1979).

Casaubon, Meric, *A Treatise Concerning Enthusiasm* (London: R.D., 1654).

——, ed., *A True and Faithful Relation of What Passed for Many Years between Dr. John Dee and Some Spirits* (London: Printed by D. Maxwell for T. Garthwait, 1659).

Chartier, Roger, *The Order of Books. Readers, Authors, and Libraries in Europe between the Fourteenth and Eighteenth Centuries*, trans. Lydia G. Cochrane (Stanford: Stanford University Press, 1994).

Cole, Peter, "The Printer to the Reader," in *The Practice of Physick ed. Lazare Rivière* (London: Printed by Peter Cole, 1655).

Collini, Stefan, *What Are Universities For?* (London: Penguin, 2012).

Cooper, Alix, "Homes and Households," in *The Cambridge History of Science, vol. 3. Early Modern Science*, eds. Katharine Park and Lorraine Daston (Cambridge: Cambridge University Press, 2006), 224–237.

Cotton, Nancy, *Women Playwrights in English, c. 1363–1750* (Lewisburg: Bucknell University Press, 1980).

Coudert, Allison, *The Impact of the Kabbalah in the Seventeenth Century* (Leiden: Brill, 1999).

Croll, Morris W., *'Attic' and Baroque Prose Style. The Anti-Ciceronian Movement*, eds. J. Max Patrick and Robert O. Evans (Princeton: Princeton University Press, 1969).

Croll, Oswald, *A Treatise of Oswaldus Crollius of Signatures of Internal Things* (1609) in *Royal and Practical Chymistry* (London: Printed for John Starkey and Thomas Passinger, 1670).

Darnton, Robert, "What Is the History of Books?" [*Daedalus* (Summer 1982), 65–83], reprinted in *The Book History Reader*, eds. David Finkelstein and Alistair McCleery (London: Routledge, second edition, 2006), 9–26.

Daston, Lorraine, "Strange Facts, Plain Facts, and the Texture of Scientific Experience in the Enlightenment," in *Proof and Persuasion. Essays on Authority, Objectivity and Evidence*, eds. Suzanne L. Marchand and Elizabeth Lunbeck (Turnhout: Brepols, 1996), 42–59.

——, "The Empire of Observation, 1699–1800," in *Histories of Scientific Observation*, eds. Lorraine Daston and Elizabeth Lunbeck (Chicago: University of Chicago Press, 2011), 81–113.

Daston, Lorraine, and Katherine Park, *Wonders and the Order of Nature* (New York: Zone Books, 2001).

Dear, Peter, "The Meanings of Experience," in *The Cambridge History of Science, vol. 3. Early Modern Science*, eds. Katharine Park and Lorraine Daston (Cambridge: Cambridge University Press, 2006), 106–131.

de Beer, E.S., "King Charles, *Fundator et Patronus* (1630–1685)," *Notes and Records of the Royal Society of London* 15 (1960), 39–45.

della Porta, Giambattista, *Natural Magick*, translated anonymously (London: Printed for Thomas Young and Samuel Speed, 1658).

Derham, William, *Physico-Theology* (London: Printed for W. Innys, 1714).

Devlin, Keith, *The End of Logic and the Search for a New Cosmology of Mind* (New York: Wiley, 1997).

Dobranski, Stephen B., *Readers and Authorship in Early Modern England* (Cambridge: Cambridge University Press, 2005).

Dolnick, Edward, *The Clockwork Universe. Isaac Newton, the Royal Society, and the Birth of the Modern World* (New York: HarperCollins, 2011).

Eisenstein, Elizabeth L., *The Printing Press as an Agent of Change. Communications and Cultural Transformations in Early Modern Europe*, 2 vols. (Cambridge: Cambridge University Press, 1979).

——, *The Printing Revolution in Early Modern Europe* (Cambridge: Cambridge University Press, second edition, 2005).

Enos, Theresa, ed., *Encyclopedia of Rhetoric and Composition. Communication from Ancient Times to the Information Age* (New York: Routledge, 1996; paperback edition, 2010).

The European Commission's Seventh Framework Programme for Research and Technological Development, http://ec.europa.eu/research/fp7/pdf/fp7-factsheets_en.pdf (accessed 28 November 2012).

Fahnestock, Jeanne, *Rhetorical Figures in Science* (New York: Oxford University Press, 1999).

Feather, John, *A History of British Publishing* (London: Routledge, second edition, 2005).

——, "The British Book Market 1600–1800," in *A Companion to the History of the Book*, eds. Simon Eliot and Jonathan Rose. Blackwell Companions to Literature and Culture, vol. 48 (Malden: Blackwell Publishing, 2007), 232–246.

Ferguson, John, "Some English Alchemical Books," *Journal of the Alchemical Society* 2 (1913), 2–6.

Feuer, Lewis, *The Scientific Intellectual* (New York: Basic Books, 1963).

Findlen, Paula, *Possessing Nature. Museums, Collecting, and Scientific Culture in Early Modern Italy*. Studies on the History of Society and Culture, vol. 20 (Berkeley: University of California Press, 1994).

Fludd, Robert, *Mosaicall Philosophy* (London: Printed for Humphrey Moseley, 1659).

Furdell, Elizabeth Lane, *Publishing and Medicine in Early Modern England* (Suffolk: Boydell and Brewer, 2002).

Gascoigne, John, "The Royal Society and the Emergence of Science as an Instrument of State Policy," *The British Journal for the History of Science* 32, 2 (1999), 171–184.

Gaukroger, Stephen, *The Emergence of a Scientific Culture. Science and the Shaping of Modernity, 1210–1685* (Oxford: Clarendon, 2006).

——, *Francis Bacon and the Transformation of Early-Modern Philosophy* (Cambridge: Cambridge University Press, 2001).

Gay, John, *The Shepherd's Week in Six Pastorals* (London: F. Burleigh, 1714).

Giere, Ronald N., "Naturalism," in *The Routledge Companion to Philosophy of Science*, eds. Martin Curd and Stathis Psillos (London: Routledge, 2008), 213–223.

Glanvill, Joseph, *Plus ultra, or, The Progress and Advancement of Knowledge Since the Days of Aristotle*, in *Collected Works of Joseph Glanvill*, facsimile editions prepared by Bernhard Fabian, vol. 4 (Hildesheim and New York: G. Olms, 1979).

——, *Saducismus triumphatus, or, Full and Plain Evidence Concerning Witches and Apparitions* (London: Printed for J. Collins and S. Lownds, 1681).

Goody, Jack, *The Interface between the Written and the Oral* (Cambridge: Cambridge University Press, 1987).

Goody, Jack, and Ian Watt, "The Consequences of Literacy," *Comparative Studies in Society and History* 5, 3 (1963), 304–345.

Gross, Alan G., *The Rhetoric of Science* (Cambridge: Harvard University Press, 1990).

——, *Starring the Text. The Place of Rhetoric in Science Studies* (Carbondale: Southern Illinois University Press, 2006).

Gross, Alan G., and Joseph E. Harmon, eds., *The Scientific Literature. A Guided Tour* (Chicago: University of Chicago Press, 2007).

Gross, Alan G., Michael Reidy, and Joseph E. Harmon, *Communicating Science. The Scientific Article from the 17th Century to the Present* (Oxford: Oxford University Press, 2002).

Habermas, Jürgen, *The Structural Transformation of the Public Sphere. An Inquiry into a Category of Bourgeois Society*, trans. Thomas Burger and Frederick Lawrence (Cambridge: Polity Press, 1989).

Harkness, Deborah, *John Dee's Conversations with Angels* (Cambridge: Cambridge University Press, 1999).

Harmon, Joseph E., and Alan G. Gross, *The Craft of Scientific Communication* (Chicago: University of Chicago Press, 2010).

Harris, Steven J., "Networks of Travel, Correspondence, and Exchange," in *The Cambridge History of Science, vol. 3. Early Modern Science*, eds. Katharine Park and Lorraine Daston (Cambridge: Cambridge University Press, 2006), 341–362.

Hawking, Stephen, *A Brief History of Time* (New York: Bantam Books, 1988).

Henderson, Felicity, "Translation Theory and Practice at the Early Royal Society," *Notes and Records of the Royal Society* 67, 2 (2013), 101–122.

Heydon, John, *Theomagia* (London: Printed by T.M. for Henry Brome and Tho. Rooks, 1663).

Howell, Wilbur S., *Logic and Rhetoric in England, 1500–1700* (New York: Russell & Russell, 1961, orig. publ. Princeton University Press, 1956).

Huff, Toby E., *The Rise of Early Modern Science. Islam, China and the West* (Cambridge: Cambridge University Press, 1993).

Hunter, Lynette, "Sisters of the Royal Society. The Circle of Katherine Jones, Lady Ranelagh," in *Women, Science and Medicine 1500–1700*, eds. Lynette Hunter and Sarah Hutton (Stroud: Sutton, 1997), 178–197.

Hunter, Michael, *Science and Society in Restoration England* (Cambridge: Cambridge University Press, 1981).

——, *Establishing the New Science* (Woodbridge: Boydell Press, 1989).

——, *The Royal Society and Its Fellows 1660–1700. The Morphology of an Early Scientific Institution* (Oxford: British Society for the History of Science, second edition, 1994).

Hunter, Lynette, and Sarah Hutton, eds., *Women, Science and Medicine 1500–1700. Mothers and Sisters of the Royal Society* (Stroud: Sutton, 1997).

Hutton, Sarah, "Jones, Katherine, Viscountess Ranelagh (1615–1691)," in *Oxford Dictionary of National Biography*, online edition, Jan 2008, http://www.oxforddnb.com/view/article/66365 (accessed 15 October 2012).

Iliffe, Rob, "Lying Wonders and Juggling Tricks. Religion, Nature, and Imposture in Early Modern England," in *Everything Connects*, eds. James E. Force and David S. Katz (Leiden: Brill, 1999), 185–211.

Jackson, Roland, "Ten Years of Public Engagement," http://www.britishscienceassociation.org/people-science-magazine/march2013/ten-years-public-engagement (accessed 20 June 2013).

Jenkinson, Matthew, *Culture and Politics at the Court of Charles II, 1660–1685*. Studies in Early Modern Cultural, Political and Social History, vol. 9 (Woodbridge: Boydell Press, 2010).

Johns, Adrian, *The Nature of the Book. Print and Knowledge in the Making* (Chicago: University of Chicago Press, 1998).

——, "Science and the Book," in *The Cambridge History of the Book in Britain, vol. 4. 1557–1695*, eds. John Barnard and D.F. McKenzie (Cambridge: Cambridge University Press, 2002), 274–303.

——, "Coffeehouses and Print Shops," in *The Cambridge History of Science*, vol. 3. *Early Modern Science*, eds. Katharine Park and Lorraine Daston (Cambridge: Cambridge University Press, 2006), 320–340.

——, *Piracy. The Intellectual Property Wars from Gutenberg to Gates* (Chicago: Chicago University Press, 2009).

Jones, Richard Foster, "Science and English Prose Style in the Third Quarter of the Seventeenth Century" [Publications of the Modern Language Association, vol. 45 (1930)], repr. in R.F. Jones, *The Seventeenth Century. Studies in the History of English Thought and Literature from Bacon to Pope* (Stanford: Stanford University Press, 1951, reissued 1965), 75–110.

Kernan, Alvin, *Printing Technology, Letters and Samuel Johnson* (Princeton: Princeton University Press, 1987).

——, ed., *What's Happened to the Humanities?* (Princeton: Princeton University Press, 1997).

Knox, John, *History of the Reformation in Scotland* (1566), ed. William Dickinson (New York: Thomas Nelson, 1949).

Kroll, Richard, *The Material World* (Baltimore: Johns Hopkins University, 1991).

Latour, Bruno, *Science in Action. How to Follow Scientists and Engineers through Society* (Cambridge: Harvard University Press, 1987).

Latour, Bruno, and Steve Woolgar, *Laboratory Life. The Construction of Scientific Facts* (Princeton: Princeton University Press, 1986).

Lead, Jane, *Revelation of Revelations* (London: A. Sowle, 1683).

——, *The Heavenly Cloud Now Breaking* (London: J.L., 1681).

Lewis, C.S., *English Literature of the Sixteenth Century, Excluding Drama* (Oxford: Oxford University Press, 1954).

Locke, John, *Essay Concerning Human Understanding*, ed. Peter Nidditch (Oxford: Clarendon, 1975).

Love, Harold, *The Culture and Commerce of Texts. Scribal Publication in Seventeenth-Century England* (Amherst: University of Massachusetts Press, 1993).

Lux, David, and Harold Cook, "Closed Circles or Open Networks? Communicating Science at a Distance during the Scientific Revolution," *History of Science* 36 (1988), 179–211.

Lyons, Martyn, *A History of Reading and Writing in the Western World* (Basingstoke: Palgrave Macmillan, 2010).

Manning, Russell and Michael Byrne, *Science and Religion in the Twenty-first Century* (Norwich: SCM Press, 2013).

McDowell, Paula, "Enlightenment Enthusiasms and the Spectacular Failure of the Philadelphian Society," *Eighteenth-Century Studies* 35 (2002), 515–533.

McKenzie, D.F., *Bibliography and the Sociology of Texts* (Cambridge: Cambridge University Press, 1999).

McKitterick, David, "University Printing at Oxford and Cambridge," in *The Cambridge History of the Book in Britain*, vol. 4. *1557–1695*, eds. John Barnard and D.F. McKenzie (Cambridge: Cambridge University Press, 2002), 189–205.

Mendelson, Sara Heller, *The Mental World of Stuart Women. Three Studies* (Amherst: University of Massachusetts Press, 1987).

Merton, Robert K., *The Sociology of Science. Theoretical and Empirical Investigations*, ed. Norman W. Storer (Chicago: University of Chicago Press, 1973).

Merton, Robert K., and Harriet Zuckerman, "Patterns of Evaluation in Science. Institutionalisation, Structure and Functions of the Referee System" [*Minerva* 9, 1 (1971), 66–100], repr. in R. Merton, *The Sociology of Science*, 460–496.

Middleton, W.E. Knowles, "What Did Charles II Call the Fellows of the Royal Society?" *Notes and Records of the Royal Society of London* 32, 1 (1977), 13–16.

More, Henry, *Enthusiasmus triumphatus, or, a Discourse of the Nature, Causes, Kinds and Cure, of Enthusiasm* (London: J. Flesher, 1656).

Murphy, Andrew, "The History of the Book in Britain, c. 1475–1800," in *The Oxford Companion to the Book*, eds. Michael Felix Suarez and H.R. Woudhuysen, 2 vols. vol. 1 (Oxford: Oxford University Press, 2010), 172–179.

The Newton Project, http://www.newtonproject.sussex.ac.uk/view/texts/normalized/NATP00006 (accessed 14 October 2012).

North, Marcy L., "Women, the Material Book and Early Printing," in *The Cambridge Companion to Early Modern Women's Writing*, ed. Laura Lunger Knoppers (Cambridge: Cambridge University Press, 2009), 68–82.

Nowotny, Helga, Peter Scott, and Michael Gibbons, "Introduction. 'Mode 2' Revisited. The New Production of Knowledge," *Minerva. A Review of Science, Learning and Policy* 41, 3 (2003), 179–194.

Nurse, Paul, "The Richard Dimbleby Lecture 2012. The New Enlightenment" (28 February 2012), http://royalsociety.org/news/2012/richard-dimbleby-lecture/ (accessed 28 November 2012).

Nussbaum, Martha C., *Not for Profit. Why Democracy Needs the Humanities* (Princeton: Princeton University Press, 2010).

Olson, David R., *The World on Paper. The Conceptual and Cognitive Implications of Writing and Reading* (Cambridge: Cambridge University Press, 1994).

Ong, Walter J., *Orality and Literacy. The Technologizing of the Word* (London: Routledge, 2002).

Ormsby-Lennon, Hugh, *Swift and the Quacks* (Lanham: Rowen & Littlefield, 2011).

Paracelsus, *Works* (London: Rich. Moon & Hen. Fletcher, 1657).

Pemberton, Henry, *A View of Sir Isaac Newton's Philosophy* (London: S. Palmer, 1728).

Pepys, Samuel, *The Diary of Samuel Pepys* (1825), eds. Robert Latham and William Matthews (London: Bell & Hyman, 1970–1983).

Pettegree, Andrew, *The Book in the Renaissance* (New Haven: Yale University Press, 2010).

Petty, William, *The Collected Works of Sir William Petty*, ed. Terrence Hutchison (New York: Routledge, 1998).

Philadelphian Society, *Theosophical Transactions* (London: Printed by the Book-sellers of London and Westminster, 1697).

Pieczka, Magda, and Oliver Escobar, "Dialogue and Science. Innovation in Policy-Making and the Discourse of Public Engagement in the UK," *Science and Public Policy* 40.1 (2013), 113–126.

Pordage, John, *Theologia mystica* (London: [s.n.], 1683).

Porter, Roy, *Enlightenment. Britain and the Creation of the Modern World* (London: Allen Lane, 2000).

Potkay, Adam, *The Fate of Eloquence in the Age of Hume* (Ithaca: Cornell University Press, 1994).

Raven, James, *The Business of Books. Booksellers and the English Book Trade, 1450–1850.* (New Haven: Yale University Press, 2007).

Read, William, *Read's True and Faithful Experiments* (London: 1705).

Rebhorn, Wayne A., ed., *Renaissance Debates on Rhetoric* (Ithaca: Cornell University Press, 2000).

Richards, Jennifer, *Rhetoric* (London: Routledge, 2008).

Rivington, Charles A., "Early Printers to the Royal Society, 1663–1708," *Notes and Records of the Royal Society* 39, 1 (1984), 1–27.

Rothkamm, Jan, *Institutio oratoria. Bacon, Descartes, Hobbes, Spinoza* (Leiden: Brill, 2009).

The Royal Society, *The Record of the Royal Society of London for the Promotion of Natural Knowledge*, 4th ed. (London: Printed for the Royal Society by Morrison & Gibb, 1940).

——, "The Public Understanding of Science" (1985), http://royalsociety.org/~/media/Royal_Society_Content/policy/publications/1985/10700.pdf (accessed 28 November 2012).

——, "Science as an Open Enterprise" (2012), http://royalsociety.org/policy/projects/science-public-enterprise/report/ (accessed 26 June 2013).

——, "Strategic Plan 2012–2017," http://royalsociety.org/about-us/governance/ (accessed 2 June 2013).

Russell, Nicholas J., *Communicating Science. Professional, Popular, Literary* (Cambridge: Cambridge University Press, 2010).

Saab, Anwar, *The Good, the True and the Beautiful. A Film about Scientific Imaging* (2012) http://insideoutimaging.wordpress.com/film-project/ (accessed 26 June 2013) http://www.youtube.com/watch?v=SR7FhzIL5Bk&feature=youtu.be.

Sarasohn, Lisa T., *The Natural Philosophy of Margaret Cavendish. Reason and Fancy during the Scientific Revolution* (Baltimore: Johns Hopkins University Press, 2010).

Schaffer, Simon, "Natural Philosophy and Public Spectacle in the Eighteenth Century," *History of Science* 21 (1983), 1–43.

Serjeantson, Richard W., "Proof and Persuasion," in *The Cambridge History of Science, vol. 3. Early Modern Science*, eds. Katharine Park and Lorraine Daston (Cambridge: Cambridge University Press, 2006), 132–175.

Shadwell, Thomas, *The Virtuoso* (1676), eds. David Stuart Rhodes and Marjorie Hope Nicolson (Lincoln: University of Nebraska Press, 1966).

Shapin, Steven, *The Scientific Revolution* (Chicago: University of Chicago Press, 1996).

——, *A Social History of Truth. Civility and Science in Seventeenth-Century England* (Chicago: University of Chicago Press, 1994).

——, "Science and the Public," in *Companion to the History of Modern Science*, eds. R.C. Olby, G.N. Cantor, J.R.R. Christie, and M.J.S. Hodge (London: Routledge, 1990), 990–1007.

——, "The House of Experiment in Seventeenth-Century England" *Isis* 79, 3 (1988), 373–404.

Shapin, Steven, and Simon Schaffer, *Leviathan and the Air-Pump. Hobbes, Boyle, and the Experimental Life* (Princeton: Princeton University Press, 1985).

Shapiro, Barbara J., *A Culture of Fact. England, 1550–1720* (Ithaca: Cornell University Press, 2000).

Small, Helen, *The Value of the Humanities* (Oxford: Oxford University Press, 2013).

Sokal, Alan, "Transgressing the Boundaries. Towards a Transformative Hermeneutics of Quantum Gravity," *Social Text* 46/47 (1996), 217–252.

Spiller, Elizabeth, *Science, Reading, and Renaissance Literature. The Art of Making Knowledge, 1580–1670* (Cambridge: Cambridge University Press, 2004).

——, ed., *Seventeenth-Century English Recipe Books. Cooking, Physic and Chirurgery in the Works of Elizabeth Talbot Grey and Aletheia Talbot Howard*. Essential Works for the Study of Early Modern Women. Series III, Part Three, vol. 3 (Burlington: Ashgate, 2008).

Sprat, Thomas, *The History of the Royal Society* (1667), eds. Jackson I. Cope and Harold Whitmore Jones (Saint Louis: Washington University Studies, 1958).

Stark, Ryan J., *Rhetoric, Science, and Magic in Seventeenth-Century England* (Washington: The Catholic University of America Press, 2009).

Stewart, Larry, *The Rise of Public Science. Rhetoric, Technology, and Natural Philosophy in Newtonian Britain, 1660–1750* (Cambridge: Cambridge University Press, 1992).

Stroup, Alice, *A Company of Scientists. Botany, Patronage, and Community at the Seventeenth-Century Parisian Royal Academy of Sciences* (Berkeley: University of California Press, 1990).

Swift, Jonathan, *Gulliver's Travels* (London: Benj. Motte, 1726).

——, *Tale of a Tub* (1704), eds. A.C. Guthkelch and D. Nichol Smith (Oxford: Clarendon, second edition, 1958).

Taylor, E. Sherwood, ed., Anonymous, "An Early Satirical Poem on the Royal Society," *Notes and Records of the Royal Society of London* 5, 1 (1947), 37–46.

Thompson, C.J.S., *Alchemy and Alchemists* (London: Courier Dover Publications, 2002).

Vickers, Brian, "The Royal Society and English Prose Style. A Reassessment," in *Rhetoric and the Pursuit of Truth. Language Change in the Seventeenth and Eighteenth Centuries*, eds. Brian Vickers and Nancy S. Struever (Los Angeles: William Andrews Clark Memorial Library, University of California, 1985), 3–76.

——, ed., *Occult and Scientific Mentalities in the Renaissance* (Cambridge: Cambridge University Press, 1984).

Walsh, Lynda, *Scientists as Prophets. A Rhetorical Genealogy* (New York: Oxford University Press, 2013).

Ward, Ned, *The Fourth Volume of the Writings of the Author of the London-Spy. Prose and Verse* (London: George Sawbridge, 1709).

Wear, Andrew, *Knowledge and Practice in English Medicine, 1550–1680* (Cambridge: Cambridge University Press, 2000).

Webster, Charles, *Magic and the Making of Modern Science* (Cambridge: Cambridge University Press, 1982).

Weedon, Alexis, "The Economics of Print," in *The Oxford Companion to the Book*, eds. Michael Felix Suarez and H.R. Woudhuysen, 2 vols. (Oxford: Oxford University Press, 2010), vol. 1, 105–106.

Wigelsworth, Jeffrey R., *Selling Science in the Age of Newton. Advertising and the Commoditization of Knowledge* (Burlington: Ashgate, 2010).

Wilkins, John, *An Essay Towards a Real Character and a Philosophical Language* (London: Printed for Sa. Gellibrand and John Martyn, 1668).

Yale, Elizabeth, "Making Lists. Social and Material Technologies for Seventeenth-Century British Natural History," in *Ways of Making and Knowing. The Material Culture of Empirical Knowledge, 1400–1850*, eds. Pamela H. Smith, Amy Meyers, and Harold Cook (Ann Arbor: University of Michigan Press, 2014).

——, "With Slips and Scraps. How Early Modern Naturalists Invented the Archive," *Book History* 12 (2009), 1–36.

PART 1

Theoretical Perspectives

CHAPTER 1

Totius in verba

Rhetoric and Authority in the Early Royal Society

*Peter Dear**

> Simplicio is confused and perplexed, and I seem to hear him say, "who would there be to settle our controversies if Aristotle were to be deposed?"
>
> – SAGREDO, in Galileo's *Dialogue Concerning the Two Chief World Systems*, trans. STILLMAN DRAKE.

THE SCIENTIFIC REVOLUTION reached a stage of consolidation in the second half of the seventeenth century. The ferment of new ideas, and the conscious rejection of old, had resulted in the formation of an identifiable community of practitioners with a shared ideal of natural inquiry. The clearest indicators of this consolidation are the scientific societies that arose in the 1650s and 1660s, foremost among them the Académie Royale des Sciences in Paris and the Royal Society of London. These groups stand as testimony to a new attitude toward knowledge of nature.[1]

This article, concentrating on the Royal Society, examines how the cooperative investigation of nature both shaped and was made possible by the new forms of natural knowledge generally associated with the Scientific Revolution. The present study, however, does not mean to suggest that these new conceptions emerged suddenly with the formation of the Royal Society, nor that they were especially English. (Some recent literature on the connections between experimental philosophy and ideological currents in Restoration England is curiously Anglocentric.)[2] Instead, the Society is treated here as a convenient focus at the end of a process of change.

---

* I would like to thank Michael Mahoney, Pauline Dear, Tom Broman, and John Carson for discussion and advice, and Steven Shapin for his encouragement.

1 On scientific societies in general, see Harcourt Brown, *Scientific Organizations of Seventeenth Century France* (Baltimore: 1934); Martha Ornstein, *The Role of Scientific Societies in the Seventeenth Century* (1913; New York: Arno Press, 1975).

2 See esp. James R. Jacob and Margaret C. Jacob, "The Anglican Origins of Modern Science. The Metaphysical Foundations of the Whig Constitution," *Isis* 71 (1980), 251–267, which claims that the connection between religious ideology and matter theory reveals the "social genesis of the conceptual revolution that culminated in the Newtonian synthesis, the hallmark of the Scientific Revolution" (254). This article uses a remarkably conventional view of the Scientific Revolution, one resting on a European perspective, to justify the broader significance of a very focused analysis of the English context.

 | DOI 10.1163/9789004283701_003

FIGURE 1.1 *Frontispiece to Thomas Sprat's* History of the Royal Society of London (*London, third edition, 1722*), *displaying the Society's motto, suggested by John Evelyn, adopted in 1662. From* "Nullius addictus iurare in verba magistri," Horace, *Epistles* 1.1.14.
SOURCE: COURTESY OF THE E.F. SMITH MEMORIAL COLLECTION, SPECIAL COLLECTIONS CENTER, UNIVERSITY OF PENNSYLVANIA.

The Royal Society has been the most intensively examined of the seventeenth-century scientific academies. Studies have been made of the scientific work carried out by Fellows of the Society and of the Society as the representative of particular social and ideological movements in Restoration England. In addition, work has been done on the epistemological positions promoted by many of the Fellows.[3] The present article considers the Royal Society from a microsocial perspective, as the embodiment of an ideal of cooperative research. Although the Society did little to control natural philosophical endeavor in a strong, programmatic, and corporate sense, as will be discussed below, its very existence and the desire of so many people to belong to it are themselves indicators of the Royal Society's significance *qua* society. The Royal Society is worth studying, not because it followed any enforced and exclusive code of practice, but because it can be taken as a symbol of particular conceptions of natural philosophy in this period especially of *cooperative* natural philosophy.

For cooperative inquiry to be conducted, it was necessary to agree upon standards and criteria; conversely, the credibility of a contribution to the common endeavor had to be determined by appeal to some sort of authority. The place of experience in the "new philosophy" played a crucial part in this process, first, because the idea of experiment and observation played an important role in the Scientific Revolution generally and in the Royal Society in particular, and, second, because polemical antischolastic writings of the period opposed experience to an alleged reliance on ancient authority. I shall argue below that not only did authority take on a new guise, but that in its doing so, the concept of experience and its place in natural knowledge also changed.

The significance of the image of the "book of nature" so frequently invoked in the sixteenth and seventeenth centuries may lie precisely in this relationship

---

3 For the scientific work of the Society see, e.g., Dorothy Stimson, *Scientists and Amateurs. A History of the Royal Society* (New York: Schuman, 1948). For social aspects, see Michael Hunter, *Science and Society in Restoration England* (Cambridge: Cambridge Univ. Press, 1981); Hunter, "The Social Basis and Changing Fortunes of an Early Scientific Institution. An Analysis of the Membership of the Royal Society, 1660–1685," *Notes and Records* 31 (1976), 9–114; Hunter, *The Royal Society and Its Fellows, 1660–1700. The Morphology of an Early Scientific Institution* (British Society for the History of Science Monographs, 4) (Chalfont St. Giles, Bucks.: British Society for the History of Science, 1982); James R. Jacob, *Robert Boyle and the English Revolution. A Study in Social and Intellectual Change* (New York: Burt Franklin, 1977); and Jacob, *Henry Stubbe, Radical Protestantism and the Early Enlightenment* (Cambridge: Cambridge Univ. Press, 1983), esp. Ch. 5. On epistemological positions see esp. Barbara Shapiro, "History and Natural History in Sixteenth- and Seventeenth-Century England. An Essay on the Relationship between Humanism and Science," in *English Scientific Virtuosi in the 16th and 17th Centuries* (Los Angeles: William Andrews Clark Memorial Library, 1979), 3–55; Shapiro, *Probability and Certainty in Seventeenth-Century England* (Princeton, N.J.: Princeton Univ. Press, 1983); and also, esp. on Boyle, Henry G. Van Leeuwen, *The Problem of Certainty in English Thought 1630–1690* (The Hague: Martinus Nijhoff, 1963).

between a new structure of authority and a new view of experience. Just as a written text needs to be read and at some level to be glossed, so the "book of nature" must be interpreted. In using the latter text, the early Fellows of the Royal Society employed new techniques of exegesis. The final product of their work was the book, letter, or article in which that work became a property of the whole community. Hence an examination of form will contribute perhaps at least as much as an examination of content to an understanding of what the new philosophy, as exemplified by the early Royal Society, actually was.

I

As the frontispiece to Thomas Sprat's *History of the Royal Society* testifies, the early Fellows liked to think of themselves as following the precepts and method of Francis Bacon. There, beside their royal patron, sat the noble *Artium Instaurator*, as if to lend his seal of approval to the Fellows' activities.[4] The Society itself, therefore, had a ready-made analogy: "*Solomon's House* in the NEW ATLANTIS, was a Prophetick Scheam of the ROYAL SOCIETY," said Glanvill; indeed, Bacon's "SOCIETY of *Experimenters* in a *Romantick Model*" had inspired the virtuosi intentionally to "set themselves on work upon this *grand* Design."[5] Certainly, such claims indicate an awareness that cooperative inquiry was a novel sort of cognitive enterprise, one demanding that knowledge be handled according to new, special criteria. But those criteria did not come from Bacon's account of Solomon's House. That notional institution divided knowledge into its component parts, with personnel assigned to distinct, autonomous tasks all contributing to the whole. The structure reflected Bacon's belief that the acquisition of knowledge was somehow an automatic process, once the correct procedure was followed.[6]

4 Thomas Sprat, *History of the Royal Society* (London, 1667); facs. eds. Jackson I. Cope and Harold Jones (Saint Louis: Washington Univ. Press, 1958), frontispiece. For the view of Bacon as a convenient emblem for the early Royal Society see, e.g., Barbara Shapiro, *John Wilkins, 1614–1672. An Intellectual Biography* (Berkeley: Univ. California Press, 1969), 205–206. Margery Purver, *The Royal Society. Concept and Creation* (Cambridge, MA: MIT Press, 1967), argues for the basic Baconianism of the early Society; but see the review by Charles Webster, "The Origins of the Royal Society," *History of Science* 6 (1967), 106–128.

5 Joseph Glanvill, "An Adress to the Royal Society," in *Scepsis scientifica* (London, 1665), facs. ed. in Glanvill, *The Vanity of Dogmatizing. The Three Versions,* intro. by Stephen Metcalf (Sussex: Harvester Press, 1970), 22; Glanvill, quoted in Purver, *Royal Society*, 96.

6 Francis Bacon, "The New Atlantis," in *The Works of Francis Bacon*, eds. J. Spedding, R.L. Ellis, and D.D. Heath, 7 vols. (London, 1859–1874), vol. III, 125–166, esp. pp. 164–165.

Despite Bacon's ideal of a cooperative investigative institution, the Royal Society in reality failed even to act as a successful coordinator of the projects of individuals. Shared ideals among the early Fellows did not, as a rule, translate into shared projects or programs of research; half-hearted attempts at promoting the compilation of Baconian histories of trades bore little or no fruit, despite Sprat's extravagant claims.[7] The Royal Society was more of a club than a college. If it did not resemble Solomon's House, however, the Royal Society did have a collective voice. The minutes of its meetings, its *imprimatur*, and, although technically run by Oldenburg in a private capacity, the *Philosophical Transactions*—all served to underline the corporate character of the Society, and hence the notion of cooperative endeavor—of what the work overseen (at least) by the official apparatus of the Society ought to be.

As a first approximation, one can describe the "research ideal" of the society as an association of individuals with common interests who were engaged in work that was commonly perceived to be of value; cooperation in the work itself was not part of this ideal. What common values did the members share? First, there had to be tacit agreement on the nature of knowledge. This, in turn, implied a common standard of authority, and ancient authority, the Fellows habitually claimed, no longer served. But what constituted ancient authority?

For Aristotle, who was to become the preeminent "ancient authority," phenomena were, literally, *data*, "givens." They were statements about how things behave in the world, and they were to be taken into account when discussing topics concerning nature. The immediate sources of phenomena were diverse: common opinion and the assertions of philosophers, as well as sense-perception. Given these statements, a system of syllogistic reasoning yielded, in principle, a theoretical description and explanation of them. This was "dialectical" rather than "demonstrative" reasoning, based on probable rather than certain grounds. Because "probable" meant "worthy of approbation" rather than simply "likely," the probability of a statement was intimately bound up with matters of authority: the more authoritative the source, the more probable the statement.[8]

---

7 See Walter E. Houghton, Jr., "The History of Trades. Its Relation to Seventeenth-Century Thought," in *Roots of Scientific Thought. A Cultural Perspective*, eds. Philip P. Wiener and Aaron Noland (New York: Basic Books, 1957), 354–381. For Sprat's comments, see Sprat, *History*, 155–157.

8 See G.E.L. Owen, "Aristotle. Method, Physics and Cosmology," in *Dictionary of Scientific Biography*, ed. Charles C. Gillespie, 16 vols. (New York: Scribner's, 1970–1980) (hereafter DSB), vol. I, 250–258, esp. 252–254; Stephen Gaukroger, *Explanatory Structures. A Study of Concepts of Explanation in Early Physics and Philosophy* (Atlantic Highlands, N.J.: Humanities

The Middle Ages and Renaissance continued this approach toward phenomena, that is, statements about how the world behaves. Since the focus of scholastic pedagogical procedure was on disputation, emphasis was given to the proper *form* of an argument. The argument's premises were almost the least interesting part.[9] The establishment of a premise usually consisted of quoting an authority for it—above all, Aristotle. For medieval and Renaissance scholastics, as for Aristotle himself, attaching the name of an authority to a statement of experiential fact rendered it probable and hence suitable for use in argument. R.W. Southern has pointed out the almost subservient role that experience played within the framework of medieval philosophical discourse; Robert Grosseteste rebuked those who would "form their own opinions from their experiments without a foundation of doctrine." Criticizing another's premises, therefore, involved demonstrating their inconsistency with other commonly accepted statements, and the essence of such statements was their generality, as in the example "bees reproduce parthenogenetically"—all bees, at all times.[10] An "experience" was thus a statement of how things are, or how they behave, and it was taken to have been originally constructed from a large number of individual sensory impressions. As Aristotle described in the *Posterior Analytics* and Buonamici, in the sixteenth century, echoed: "Great is the power of experience which arises from the memory of things which sense time and again supplies; for indeed memory comes from repeated sensation. Many memories of the same thing grant the means of one experience."[11] But one did not need to have acquired such experiences directly in order to use them in argumentation, provided that they could be drawn from the statements of a weighty authority, and thereby rendered probable.

Press, 1978), 91–92. On premodern concepts of probability, see Ian Hacking, *The Emergence of Probability* (Cambridge: Cambridge Univ. Press, 1975), Chs. 1–5.

9 On scholasticism as a pedagogical "method," and the role of disputation, see George Makdisi, "The Scholastic Method in Medieval Education. An Inquiry into Its Origins in Law and Theology," *Speculum* 49 (1974), 640–661, esp. pp. 642–648.

10 The example comes from Aristotle, *De animalibus historiae*, 9.42. See R.W. Southern, "Commentary on Medieval Science," in *Scientific Change*, ed. Alistair C. Crombie (New York: Basic Books, 1963), 301–306, on p. 305, and for a comparison of Grosseteste's concept of "experimentum" to "experience" in the Aristotelian sense, see Bruce S. Eastwood, "Medieval Empiricism. The Case of Grosseteste's Optics," *Speculum* 43 (1968), 306–321, esp. p. 321. See also James McEvoy, *The Philosophy of Robert Grosseteste* (Oxford: Clarendon Press, 1983), esp. pp. 206–211.

11 Quoted in Charles B. Schmitt, "Experience and Experiment. A Comparision of Zabarella's View with Galileo's in *De motu*," *Studies in the Renaissance* 16 (1969), 80–138, on pp. 90–91, deriving from Aristotle, *Posterior Analytics*, 2.19.

Jean Buridan, for example, when discussing whether or not the earth moves, remarked that "the last appearance that Aristotle notes is more demonstrative [than the previous] to the matter in hand. This is that an arrow projected from a bow directly upward falls again in the same spot of the earth from which it was projected. This would not be so if the earth were moved with such velocity." The "experience" here is a general statement about how things habitually behave; furthermore, like a purely rational argument, it is ascribed to Aristotle. Buridan does not regard the point as conclusive, and he marshals counterarguments to the interpretation of the experience as proof of the earth's stability, but he does not question the truth of the experience itself.[12]

It was this subordination of experiential statements to the structure of argument, without subjecting them to much investigation, which prompted the seventeenth-century charges of slavish adherence to ancient authority, rather than any widespread and genuinely uncritical acceptance of Aristotelian texts. We can see this in the famous passage concerning the motion of bodies on a moving earth in Galileo's *Dialogue Concerning the Two Chief World Systems.* When Simplicio, defending the claim that a ball dropped from the mast of a moving ship falls aft of the mast's foot, is asked, "Have you ever made this experiment of the ship?" he is made to reply: "I have never made it, but I certainly believe that the authorities who adduced it had carefully observed it. Besides, the cause of the difference [between the effects on a moving and a stationary ship] is so exactly known that there is no room for doubt." The subsumption of experience by argumentative structure is here quite evident; Galileo's spokesman, Salviati, uses it as the basis of his criticisms.[13]

The "ancient authority" against which the seventeenth-century proponents of new approaches to natural philosophy argued was located in authoritative *texts.* Those texts laid down the accepted framework of discourse and interpretation, and this framework, supported by the position of the commentary as the standard philosophical genre, tended to come from Aristotle (or, in the case of medicine, Galen). Because the texts determined the character and function of statements of general experience, the authoritativeness of such statements derived from the place they held in the text, and the place the text held in natural philosophical inquiry, rather than from a blind faith in the assertions of Aristotle. The way in which an authoritative text could inform the

---

12 Jean Buridan, "Questions on the Four Books on the Heavens and the World of Aristotle," in *The Science of Mechanics in the Middle Ages*, ed. Marshall Clagett (Madison: Univ. Wisconsin Press, 1961), 596.

13 Galileo Galilei, *Dialogue Concerning the Two Chief World Systems*, trans. Stillman Drake (Berkeley: Univ. California Press, 1967), 144.

structure of natural philosophical discourse even when the genre of commentary was not chosen can be seen in an example from Albertus Magnus. Albertus, discussing trees, says that his procedure will be to "consider first the whole diversity of the parts": "First, however, we shall only cite these differences and afterward assign the causes of all the differences. If we did not follow Aristotle, however, but others, we would surely proceed otherwise. We say, therefore, with Aristotle, that certain plants called trees have gum—as the pine tree, resin and almond gum, myrrh, frankincense, and gum arabic." In effect, Aristotle was an "authority" because he was the author of texts used habitually as loci for the discussion of particular subjects. If one wished, for example, to write about the nature of the heavens, the appropriate procedure was to compose a commentary on *De caelo.* Within that framework, "experience" had a particular meaning and role.[14]

In rejecting ancient authority, therefore, the "moderns" were not rejecting the ancients themselves, but the role their writings played in intellectual inquiry. Again, Galileo remarked that "it is the followers of Aristotle who have crowned him with authority, not he who has usurped or appropriated it to himself." The blind adherence of the schoolmen (so the argument ran) was at

14 Albertus Magnus, *On Plants*, 1.2, trans. in *A Source Book in Medieval Science*, ed. Edward Grant (Cambridge, Mass.: Harvard Univ. Press. 1974), 692 (the text Albertus used was not, in fact, by Aristotle, although it was widely believed to be at the time). It was not just the assimilation of ancient, particularly Aristotelian, texts in the thirteenth century that promoted the commentary as a major scholarly form; the dominance of various forms of "Aristotelianisms" well into the Renaissance meant the continuing dominance of commentaries on Aristotle. See. e.g., Charles B. Schmitt, *A Critical Survey and Bibliography of Studies on Renaissance Aristotelianism 1958–1969* (Padova: Editrice Antenore, 1971); Schmitt: "Toward a Reassessment of Renaissance Aristotelianism," *History of Science* 11 (1973), 159–193; and Schmitt, *Aristotle in the Renaissance* (Cambridge, MA: Harvard Univ. Press, 1983). For a short discussion of medieval commentary see Neal W. Gilbert, *Renaissance Concepts of Method* (New York: Columbia Univ. Press, 1960), 27–31. For a description of various genres of commentary, see C.H. Lohr, "Renaissance Latin Aristotle Commentaries. Authors A-B," *Studies in the Renaissance* 21 (1974), 228–289, on pp. 228–233. Edward Grant, in "Aristotelianism and the Longevity of the Medieval World View," *History of Science* 16 (1978), 93–106, attributes the survival of the Aristotelian world view well into the seventeenth century to the status of the commentary as the prime vehicle of natural philosophical discussion. See also Ian Maclean, "The Interpretation of Natural Signs. Cardano's *De subtilitate* versus Scaliger's *Exercitationes*," in *Occult and Scientific Mentalities in the Renaissance*, ed. Brian Vickers (Cambridge: Cambridge Univ. Press, 1984), 231–252, esp. pp. 239–240. A curious echo of this genre of natural-philosophical writing is Samuel Clarke's edition of Jacques Rohault's textbook, *Physica* (London, 1697), wherein Rohault's Cartesian text is glossed with lengthy dissenting Newtonian notes by Clarke.

fault; there was nothing especially authoritative about The Philosopher. John Wilkins, later in the century, put it this way: "And though we do very much honour *Aristotle* for his profound judgment and universall learning, yet are we so farre from being tyed up to his opinions, that persons of all conditions amongst us take liberty to discent from him, and to declare against him, according as any contrary evidence doth engage them, being ready to follow the Banner of truth by whomsoever it shall be lifted up."[15]

Whether or not one questions the sincerity of such pronouncements (and there seems little reason to do so), it remains clear that Aristotle's image had changed. He was no longer The Philosopher; he was, like Socrates, a man, and he had been shackled by the state of learning in his own time. Consequently, although he had ceased to be an Authority (one would not wish any more to write commentaries on his works in order to elucidate nature), Aristotle was not himself to be reviled, for he certainly had been learned, as Henry Oldenburg, the Royal Society's secretary, acknowledged: "We say heartily, Read *Aristotle*, read him in his own Stile; read him entirely and fully; not feeding onely on his Ulcers and Excrescensies. ...chuse his best Vertues, examine and weigh all his Mathematical Illustrations, descend to his particulars." And after this indulgence, Oldenburg urges, the seeker after truth should "hasten to our *Christian* Philosophers, and they will forth-with acquaint you with the true Works and wonderful Contrivances of the Supreme Author."[16] The "Supreme Author," he suggests here, has replaced the Authority, Aristotle. The natural philosopher's proper task is to write a commentary on nature, not on the works of the Stagirite. Oldenburg's recommendation that Aristotle should be read "in his own Stile," that is, in classical Greek, serves to emphasize Aristotle's new status as a historical figure and his corresponding irrelevance to natural philosophy.

Although the people we have just quoted agreed that the authority of the ancients should be rejected, they did not agree on what should replace it. Bacon, too, had claimed that, despite his innovations, the "honour of the ancient authors remains untouched, since the comparison I challenge is not of wits or faculties, but of ways and methods";[17] nonetheless, he and Galileo would scarcely have seen eye-to-eye on most substantive questions concerning knowledge of the natural world. Similar philosophical differences existed

---

15 Galileo, *Dialogue*, 111; John Wilkins, *Vindiciae academiarum* (Oxford, 1654), facs. in Allen G. Debus, *Science and Education in the Seventeenth Century* (London: Macdonald, 1970), 196.

16 *Philosophical Transactions*, 2 (1666), 413.

17 Francis Bacon, *The New Organon*, Aphorism 32, trans. James Spedding (Indianapolis: Bobbs-Merrill, 1960), 46.

between members of the early Royal Society. William Petty, for example, believed that the world was made up of atoms; furthermore, as he declared to his fellow academicians, there were reasons for postulating the existence of two types of atom: "Male and Female created He them." After dabbling in Cartesianism, Henry More, also a Fellow of the Royal Society, subscribed to the idea of an "hylarchick spirit" which acted as an active principle in nature. Henry Power saw himself as a Cartesian, while to Newton, of course, such a position was anathema.[18]

Such deep-rooted disagreements formed no basis for uniting the membership of a philosophical society, and collective distrust of older patterns of authority scarcely constituted an adequate basis. Yet the Royal Society prospered. A new source of authority had in fact come into being, a means of facilitating a cooperative enterprise dedicated to the expansion of natural knowledge. This new authority overrode individual differences in ontological commitment.

II

Thomas Sprat, in a passage verifiable by means of a brief perusal of the *Philosophical Transactions*, emphasized the diversity of those contributing to the Royal Society's work: "We find many Noble Rarities to be every day given in, not onely by the hands of Learned and profess'd Philosophers; but from the Shops of *Mechanicks*; from the Voyages of *Merchants*; from the Ploughs of *Husbandmen*; from the Sports, the Fishponds, the Parks, the Gardens of *Gentlemen*." He continued: "If we only requir'd perfect *Philosophers*, to manage this employment, it were another case. For then I grant it were improbable, that threescore, or an hundred such should meet in one time." All that is required are "plain, diligent and laborious observers."[19] All could, in principle,

18 See Robert Kargon, "William Petty's Mechanical Philosophy," *Isis* 56 (1965), 63–66, on p. 65; William H. Austin, "Henry More," DSB, vol. IX, 509–510 (for a strong reaction against More's ideas on spiritual agencies by another Fellow of the Royal Society, see Robert Hooke, "Lampas," in *Early Science in Oxford*, ed. R.T. Gunther, 14 vols. (Oxford, 1923–1945), vol. VIII, 154–208, on pp. 182–195); and, Marie Boas Hall's introduction to Henry Power, *Experimental Philosophy* (London, 1664), facs. ed. (New York: Johnson Reprint, 1966); and Charles Webster, "The Discovery of Boyle's Law and the Concept of the Elasticity of the Air," *Archive for the History of the Exact Sciences* 2 (1965), 441–502, esp. p. 460 ff. As for Newton, all of *Principia*, Bk II, is openly anti-Cartesian.

19 Sprat, *History*, 72.

participate in the Society's activities. But in order to do this, they needed to conform to certain standards so that they could each assume the mantle of a new kind of authority.

When a Fellow of the Royal Society made a contribution to knowledge, he did so by reporting an experience. That experience differed in important respects from the definition informing scholastic practice; rather than being a generalized statement about how some aspect of the world *behaves*, it was instead a report of how, in one instance, the world had *behaved*. Consider, for example, Robert Boyle on an air-pump experiment:

> But when after this, the feathers being placed as before, we repeated the experiment by carefully pumping out the air, neither I nor any of the bystanders could perceive anything of turning in the descent of the feathers; and yet for further security we let them fall twice more in the unexhausted receiver, and found them to turn in falling as before; whereas when we did a third time let them fall in the well exhausted receiver, they fell after the same manner as they had done formerly.[20]

An "experience" was now, it appears, an event of which the observer was a part. Boyle's report conveys the impression of an actual, discrete event and of the observer's central role in it, not only by his careful recounting of the facts, but also by his use of the first person, active voice. This form of presentation is overwhelmingly the rule in the writings of the Fellows, regardless of the subject matter. Henry Power, discussing his microscopical examinations of the "sycamore locust," said: "I could not only see its eyes, which are red, goggled and prominent; but also I could see them perfectly latticed." Christopher Merrett describing his work on horticulture, reported: "*Anno* 1665, I made the following Experiment with 3 *May-Cherry Trees*."[21]

---

20 Quoted in Marie Boas Hall, *Robert Boyle on Natural Philosophy* (Bloomington: Indiana Univ. Press, 1966), 349. Steven Shapin, "Pump and Circumstance. Robert Boyle's Literary Technology," *Social Studies of Science* 14 (1984), 481–519, discusses the function of Boyle's use of detailed accounts as a form of "virtual witnessing" designed to establish "matters of fact." The importance of the latter as a form of social boundary marker is discussed in Simon Schaffer, "Making Certain" (essay review of Shapiro, *Probability and Certainty*, cit. n. 4), *Social Studies of Science* 14 (1984), 137–152. See also Steven Shapin and Simon Schaffer, *Leviathan and the Air-Pump. Hobbes, Boyle and the Experimental Life* (Princeton, N.J.: Princeton Univ. Press, forthcoming), Ch. 4.

21 Power, *Experimental Philosophy*, 32; report from Christopher Merrett, *Philosophical Transactions* 2 (1666), 455.

Accounts of other people's experiences were cast in the same way as those reported at first hand: the veracity of the report clearly depended on the original experience of a specified person on a particular occasion. No hearsay evidence could be admissible, as the following excerpt from the Royal Society's *Philosophical Transactions* reports:

> There was very lately produced a Paper, containing some observations, made by Mr. *Hook*, about the Planet Mars; in the Face whereof he affirmed to have discovered, in the late months of *February* and *March*, that there are several *Maculae* or *Spotted parts*, changing their place, and not returning to the same *Position*, till the next ensuing night near about the same time. ...This short intimation of it, is intended onely to invite others, that have opportunity, timely to make Observations, (either to confirm, or rectify) before *Mars* gets out of sight.[22]

The actuality of a discrete event was the central point to be established in any contribution to the cooperative philosophy of the Royal Society, and the tokens of good faith employed by a virtuoso, that is, the circumstantial details, often ran to excruciating length. Robert Boyle is notorious in this respect.

> We took a slender and very curiously blown cylinder of glass, of nearly three foot in length, and whose bore had in diameter a quarter of an inch, wanting a hair's breadth: this pipe, being hermetically sealed at one end was, at the other, filled with quicksilver, care being taken in the filling, that as few bubbles as was possible should be left in the mercury. Then the tube being stopt with the finger and inverted, was opened, according to the manner of the experiment, into a somewhat long and slender cylindrical box (instead of which we are now wont to use a glass of the same form) half filled with quicksilver: and so, the liquid metal being suffered to subside, and a piece of paper being pasted on level with its upper surface, the box and tube and all were by strings carefully let down into the receiver.[23]

Much of this last passage is written in the passive voice. This practice was not uncommon in those sections of a report dealing with the setting-up of an

22 *Philosophical Transactions* 1 (1666), 198. I am not aware of Hooke's apparent observation of the axial rotation of Mars having been noted previously; Cassini had first reported it earlier that same year (1666); see Rene Taton, "Cassini I," *DSB*, vol. III, 100–104, on p. 101.

23 Quoted in Hall, *Robert Boyle*, 329–330.

experiment; the experimenter would abandon it when the "experience" itself came to be described (as we saw in the previous quotation from Boyle). Frequently, a recipe-like format appeared, the reader being provided with a set of instructions for experimental procedure. Again, the writer would switch to a description of what had happened, once the preparations for the experience were out of the way.[24] The procedure could always be repeated; the event could never be. When a contribution to the *Philosophical Transactions* by Richard Lower breached the usual protocol, the editorial hand of Henry Oldenburg apparently intervened. The piece, concerning a way of "*making a* Dog *draw his Breath exactly like a* Wind-broken Horse," explained the technique involved by means of instructions. However, the outcome was improperly presented: "You will plainly see him draw his breath exactly like a *Wind-broken Horse*." The absence of a discrete experience as the culmination of the report appears to have deprived it of proper credentials; accordingly, the deficiency is compensated for by these words, which preface the article: "This Experiment was made before the *Royal Society*, Octob. 17. 1667 after it had been tried by the Author in private some while before."[25] An experience was linked to a particular time and place.

Examples can be chosen at random to show how the presentation of experiments or observations (both "experiences") served to place the reporter at the center of an event. The described experience was *discrete*; it was a single, historical occurrence, not a generalized statement. These things, we seem to be told, had happened by the action of or in the presence of a particular person, at a particular time and place. The specificity and consequent verisimilitude of the presentation lent the described experience an authority functionally equivalent to, but different from, that deriving from the use of an authoritative ancient text. With the rejection of ancient authority, a new way of supporting one's statements was needed; no longer were detailed references to Aristotle acceptable as premises for assertions or as frames within which to construct arguments. The new approach to knowledge entailed the rooting of claims about the natural world in discrete events. Boyle criticized those who "neither

24 See, e.g., "An Extract of a Letter Lately Written by an Observing Friend of the *Publisher*, Concerning the Vertue of *Antimony*," *Philosophical Transactions* 3 (1668), 774. The experience proceeds thus: "I tried that a Boare, to whom I had given an ounce of crude Antimony at a time, puting him into the Sty, would be fat a fortnight before another, having no Antimony, upon the like feeding…" The "recipe" section begins: "The manner of using it, is this. Take one drachme of crude Antimony powder'd for one Horse, and when you given him his Oats in a morning…"

25 *Philosophical Transactions* 2 (1667), 544–546, on p. 545, 544.

themselves ever took the pains to make trial of [experiments], nor received from any credible persons that professed themselves to have tried them";[26] for knowledge about the world was to be provided by a definite occurrence happening to a particular person. The resulting style of presentation allowed no clear distinction to be made between a "natural historical" (in the modern sense) and an "experimental" report; each was, in the same way, given as an experience defined in space and time by an actor, the observer. The credentials that established the actuality of the event were provided by surrounding the description by a wealth of circumstantial detail. This detail generally included information regarding time, place, and participants, together with additional extraneous remarks about the experience, all serving to add verisimilitude. (John Beale, in one contribution, was careful to note that a mackerel left to rot so as to exhibit phosphorescence had been prepared by his cook.[27]) Frequent use of the active voice served to strengthen the impression of an actual event and the observer's central role within it. All of this constituted the report of an experience, and the literary form itself represented a credential formerly supplied by references to an appropriate ancient, or by a different literary form-the commentary. The ancients themselves suffered a change of image consonant with the new authority. The microscope, said Henry Power, had revealed a whole new world of living creatures, "so that were *Aristotle* now alive, he might write a new History of Animals."[28]

That the new conception of authority had become a virtual *sine qua non* of the experimental philosophy is evident in the paper sent by Isaac Newton to the Royal Society in 1672. Newton had taken pains to cast the piece in the correct mold, so that it should carry the proper weight. What is interesting is that the scenario he describes, the experience he records, is a fabrication.

> To perform my late promise to you, I shall without further ceremony acquaint you, that in the beginning of the Year 1666 (at which time I applyed myself to the grinding of Optick glasses of other figures than Spherical,) I procured me a Triangular glass-Prisme, to try therewith the celebrated *Phaenomena* of *Colours*. And in order thereto having darkened my chamber, and made a small hole in my window-shuts, to let in a

26 Quoted in Hall, *Robert Boyle*, 131.

27 "An Experiment to Examine, What *Figure*, and *Celerity* of *Motion* begetteth, or encreaseth *Light* and *Flame*," *Philosophical Transactions* 1 (1666), 226–228.

28 Power, *Experimental Philosophy*, preface (unpaginated), [7]–[8].

> convenient quantity of the Suns light, I placed my Prisme at his entrance, that it might be thereby refracted to the opposite wall.[29]

This famous account, it has been amply shown, was a spurious description of the origin of Newton's ideas on light. Newton's research was conducted over a period of some years, and his presentation to the Royal Society is an idealized version representing a whole series of investigations.[30] He gave the Society just the sort of thing which it required—an event, in which he was the central participant. When delivering his optical ideas to those undergraduates (if any) who attended his *lectiones opticae* at Cambridge in 1670, however, Newton adopted a completely different strategy, one designed to meet the demands of a different forum, a different approach to knowledge. Instead of an event, Newton provided a demonstration, and he gave his conclusion at the beginning, as a postulate to be proved, rather than at the end, as a deduction from a discrete series of experiments. After having stated baldly that "Rays in respect to the Quantity of Refraction differ from one another," Newton presented a geometrically defined exposition of that assertion: "For a farther Illustration of this, let EFG (in *Fig.* 1) be any refracting surface, suppose of Glass, and let any Line OF be drawn meeting it in F." Following this precise, formal enunciation, Newton proceeded with the proof: "Our Opinion in this Matter being thus briefly explained, that you may not think we have declared to you Fables instead of Truth, we shall immediately produce the Reasons and Experiments on which these things are founded." Then came the prism experiment—but not in the form of an historical narration: "Let (in *Fig.* 2) F be any Hole in the Wall or Shut of a Chamber Window, through which the solar Rays OF may be transmitted, all other Avenues being every where carefully closed, lest the Light may enter elsewhere. ...Then let be applied to that Hole the triangular glass Prism AaBbCc." The difference between Newton's presentation in the Cambridge lectures and his presentation in the letter to the Royal Society

29 Isaac Newton, "New Theory about *Light* and *Colors*," *Philosophical Transactions* 6 (1672), 3075–3087; on pp. 3075–3076.

30 See, e.g., Richard S. Westfall, "The Development of Newton's Theory of Color," *Isis* 53 (1962), 339–358, esp. pp. 351–352; Westfall, *Never at Rest. A Biography of Isaac Newton* (Cambridge: Cambridge Univ. Press, 1980), 156–174, 211–222, esp. pp. 156–164; Thomas S. Kuhn, "Newton's Optical Papers," in *Isaac Newton's Papers and Letters on Natural Philosophy*, ed. I. Bernard Cohen (Cambridge, MA: Harvard Univ. Press, 1958), 27–45, esp. pp. 33–34, n. 11: "The implication of Newton's account of 1672 is wrong in that Newton did not proceed so directly or so immediately from the first prism experiment to the final version of the theory as the first paper would imply."

demonstrates that the Society (as did Cambridge) embodied a particular approach to natural philosophy. Cooperative experimental research required that Newton construct a discrete event as the foundation for his claims—not as a cynical attempt to gain credibility, but as a matter of propriety.[31]

In at least one instance, then, the event described did not actually take place. But since part of the purpose of the rhetorical form we have examined was to establish that the event did take place, a further factor, that is, the identity and status of those involved, clearly held some importance. We have seen Sprat boasting of the wide diversity of people who contributed to the Royal Society's project, but it should not be imagined that all were therefore equal. The humble could play their part, but not in quite the same way as the mighty. Gentility certainly added weight in its own right: "To the truth of these Relations [concerning two elderly people who suddenly sprouted new teeth], not onely the said *Joseph Shute* and *Maria Stert* [the recipients of the teeth], have put the one his name, the other her Mark, the third and seventh of *January*, 1666, but also Sir *William Strode*, and Mr. Colepresse have subscribed the same, as believing the Relation to be true." Nobility sometimes served to reduce the required amount of circumstantial description needed for accreditation. "This story Sir R. *Moray* affirmed to have received from the *Earl* of *Weymes*, Brother in Law to the Lord *Sinclair*, as it was written to him from *Scotland*."[32] And Boyle's credibility was probably not unconnected to the fact that he was "doubly Honourable (both for his parts and parentage)"[33]—though in his case details did not thereby suffer. The Society was very conscious of its social position; as Sprat's listing shows, among the Fellows were included a great number of churchmen and aristocrats who had no real connection with experimental philosophy.[34] The Royal Society as a group thereby gained not only some small measure of additional financial patronage, but also social prestige, which could itself be turned to evidential advantage. Hooke, in order to lend force to a purported observation, remarked that he had "found by a most certain Experiment, which

---

31 Isaac Newton, *Optical Lectures Read in the Publick Schools of the University of Cambridge* (London, 1728), 5, 7; on the strategy see Kuhn, "Newton's Optical Papers," 32, n. 10, 35. The formal quality of Newton's lectures is consonant with their being originally delivered in Latin, unlike his letter in English to the Royal Society. *Optical Lectures*, a translation of the original "lectiones opticae," has recently been edited by Alan Shapiro in *The Optical Papers of Isaac Newton, vol. I. The Optical Lectures, 1670–72* (Cambridge: Cambridge Univ. Press, 1984).

32 *Philosophical Transactions* 1 (1666), 381, 45.

33 Power, *Experimental Philosophy*, preface, [18].

34 See Sprat, *History*, 431–433, for a list of members, and Hunter, *Fellows of the Royal Society*, for a full analysis.

I exhibited before divers illustrious Persons of the *Royal Society*, that the Refraction of Water was greater than that of Ice, though some considerable Authors have affirm'd the contrary."[35]

At the other end of the scale were the humble gardeners and merchants: "Here follows a Relation…which is about the new *Whale-fishing* in the *West-Indies* about the *Bermudas*, as it was delivered by an understanding & hardy Sea-man, who affirmed to have been at the killing work himself." Gentlemen were trustworthy just because they *were* gentlemen; lowly folk possessed a different asset. Boyle passed on a paper concerning ambergris with the following comment: "And probably you will be invited to look on this account, though not as complete, yet as very sincere, and on that score Credible, if you consider, that this was not written by a Philosopher to broach a *Paradox*, or serve an *Hypothesis*, but by a Merchant or Factor for his Superiors, to give them an account of a matter of fact."[36] Everyman had one advantage his betters could so easily lack—he was unlikely to be prejudiced by hypotheses. The importance of this trait shows from judgments made of those who might be so prejudiced. A review in *Philosophical Transactions* of a book by Hevelius not only stressed Hevelius's attempts to show what "care and diligence" he had employed in his observations of a comet, but also noted that these qualities showed that he had not been "prepossessed by any *Hypothesis*."[37] Boyle's merchant was *a priori* less suspect on this score, and that lent authority to his assertions.

III

The Royal Society's empiricism was rooted in the authority of the individual reporter as the actor in a well-defined, particular experience. The strict limitation that this created on claims to knowledge was essential to the successful functioning of the Society; the various "theoretical" positions held by its members were far too diverse for a consensus to be possible except in a very general way. Sprat expressed it in terms of avoiding error: the Society's work ought to be entrusted "not to a *Company* all of *one mind*"; the "heat of invention" should not lead the experimental philosopher to "swallow a deceit too soon."[38] Whatever Sprat's own reasons for stressing the point, this also happened to be

35 Robert Hooke, *Micrographia* (London, 1665), 220; on patronage see Shapiro, *John Wilkins*, 194–195.

36 *Philosophical Transactions* 1 (1665), 11; *ibid.*, 7–8 (1673), 6113.

37 *Ibid.*, 1 (1666), 303.

38 Sprat, *History*, 73.

the common view of the Fellows. Henry Power approvingly quoted Boyle on the matter:

> When a writer, saith he, acquaints me onely with his own thoughts or conjectures, without inriching his discourse with any real Experiment or Observation, if he be mistaken in his Ratiotination, I am in some danger of erring with him, and at least am like to lose my time, without receiving any valuable compensation for so great a loss: But if a Writer endevours, by delivering new and real Observations or Experiments, to credit his Opinions, the case is much otherwayes; for, let his Opinions be never so false (his Experiments being true) I am not obliged to believe the former, and am left at my liberty to benefit my self by the latter.[39]

One problem with hypotheses, of course, was that they could not be grounded in the bedrock of the new authority. The speculations of one person no longer found acceptable justification in ancient precedent, nor could they by way of compensation assume the status conferred by the device of participant and discrete experience. Hypotheses remained a matter of choice, at best of heuristic, and could play no active part in the furtherance of the experimental philosophy that was the basis of the Royal Society's cooperative scheme.[40] Outside the limits of the Society's activities, however, the individual natural philosopher could quite properly utilize hypotheses. Hooke's well-known disclaimer at the beginning of his *Micrographia* makes just this point. Praising the Society for "avoiding *Dogmatizing*, and the espousal of any *Hypothesis* not sufficiently grounded and confirm'd by *Experiments*," he apologizes for any "*Expressions*, which may seem more *positive* then YOUR Prescriptions will permit"; if he has gone too far, "'tis fit that I should declare, that it was not done by YOUR Directions." Theorizing was not in keeping with the Society's function, but it was not ruled out *tout court*.[41]

---

39 Power, *Experimental Philosophy*, preface, [18]–[19].

40 For a good treatment of hypothesis in the early Royal Society, see Marie Boas Hall, "Science in the Early Royal Society," in *The Emergence of Science in Western Europe*, ed. Maurice Crosland (New York: Science History Publications, 1976), 57–77; also Shapiro, *Probability and Certainty*, 44–61.

41 Hooke, "To the Royal Society," preface to *Micrographia*. Barbara Shapiro, in her examinations of this question, does not make a clear distinction between the propriety of hypothesis for individual investigation and its suspect character within the common forum of debate. E.g.: "The turning point in England seems to have come somewhere about 1650, for from the 1650s to Newton's hostile pronouncements at the end of the century, a substantial portion of English scientists found theory, conjecture, and hypothesis extremely

If a Fellow wished some of his theoretically informed research to be added to the Royal Society's work, then, he had to elevate its status above that of a conceit and provide it with all the appropriate trappings of authority. In the third number of the *Philosophical Transactions* is a fairly lengthy report concerning "*Some Observations and Experiments upon May-Dew*" communicated by "That ingenious and inquisitive Gentleman, Master *Thomas Henshaw*." K. Theodore Hoppen has pointed out that May dew had important alchemical significance, thereby explaining Master Henshaw's great concern with such a surprising substance. None of this, however, is even hinted at in the *Philosophical Transactions* report. Instead, one reads, in the usual style: "He observed, that at the beginning, within twenty-four hours, a slimy film floated on the top of the water"; "He found the glass almost filled with an innumerable Company of small Flyes"; and so forth.[42] The theoretical justification for the experiments could not properly be presented; what took its place was a correctly accredited experience.

John Wallis's theory of the tides, read to the Society in 1666 and printed in the *Philosophical Transactions*, presented similar problems. Precisely because it was a theory (being a modification of Galileo's, adjusted so as to include and account for monthly periodicity), it conformed rather uncomfortably to the format of the Royal Society.[43] However, one consequence of Wallis's model was that the highest tides should occur not at the equinoxes, as was popularly held to be the case, but during February and November. This circumstance provided him with an opportunity not only to seek confirmation for his theory, but also to justify presenting it by reporting experiences. After giving the testimony of the people of Romney Marsh, which he had heard before he had ever

useful" ("History and Natural History," 32), a position upheld also in *Probability and Certainty*, esp. Ch. 2. Aside from asking who in England before 1650 (apart from Bacon) represented a strongly contrasting empiricism, one should bear firmly in mind the strongly circumscribed place of hypothesis within the auspices of the Royal Society. This social rather than epistemological requirement sprang from the Fellows' recognition of its importance in that particular forum rather than from any rigid enforcement by a corporate apparatus.

42 *Philosophical Transactions* 1 (1665), 33–36, quoting 33, 34, 35. On Henshaw, see Stephen Pasmore, "Thomas Henshaw, F.R.S. (1618–1700)," *Notes and Records* 36 (1982), 177–188; on Maydew see K. Theodore Hoppen, "The Nature of the Early Royal Society," *British Journal for the History of Science* 9 (1976), 246.

43 See Thomas Birch, *The History of the Royal Society of London* (London, 1756–1757), vol. I, 88–89; *Philosophical Transactions* 1 (1666), 263–289, esp. 271–275. For a brief secondary account, see E.J. Aiton, "Galileo's Theory of the Tides," *Annals of Science* 10 (1954), 44–57, on pp. 50–54.

considered questioning the usual opinion, Wallis proceeded to bolster his claim in the correct fashion:

> And since that time, I have my self very frequently observed (both at *London* and elsewhere, as I have had occasion) that in those months of *February* and *November*, (especially *November*) the Tides have run much higher, than those at other times: Though I confess, I have not been so diligent to set down those Observations, as I should have done. Yet this I do particularly very well remember, that in *November* 1660 (the same year that his Majesty returned) having occasion to go by Coach from the *Strand* to *Westminster*, I found the Water so high in the middle of King-street, that it came up, not onely to the Boots, but into the body of the Coach; and the *Pallace-yard* (all save a little place near the *West-End*) overflow'd; as likewise the Market-place; and many other places; and their Cellars generally filed up with Water.[44]

Rather like Newton, though in a less thoroughgoing way, Wallis tried to increase the acceptability of an hypothesis by adhering as much as possible to the proper evidential form.[45]

IV

The style of science espoused by the Fellows of the Royal Society was more important than the substance of that science. The form of their research reports, as well as prefatory and programmatic statements, demonstrates an ethic of investigation suitable to the ideal of cooperative research. This form also indicates that a fundamental change in concepts of experience and authority in natural philosophy had occurred in the seventeenth century, a change that underlay contemporary charges of scholastic vassalage to ancient authority. Ancient authority was rooted in the use of the writings of, above all, Aristotle as the proper vehicles of philosophical discussion, and experience appeared in the form of a generalized statement of fact, its role dictated by the overall structure of discourse. By contrast, when a Fellow of the Royal Society

44 *Philosophical Transactions* 1 (1666), 275–276.

45 See esp. Zev Bechler, "Newton's 1672 Optical Controversies. A Study in the Grammar of Scientific Dissent," in *The Interaction between Science and Philosophy*, ed. Yehuda Elkana (Atlantic Highlands, N.J.: Humanities Press, 1974), 115–142; Newton presents his hypothesis as if it were not a hypothesis at all.

utilized experience, he provided a report of a discrete event, independent of the setting of an ancient text or conceptualization.[46] In such reports, the operator or observer (the two were equivalent) was central to the episode recounted—and episodes they were. Located, explicitly or implicitly, at a precise point in space and time, the observer's reported experience of a singular phenomenon constituted his authority. Newton resorted to fabricating an event as a means of conferring this authority on his optical theories.

This article has touched only briefly on mathematical sciences. There are, in fact, very few mathematical contributions to be found in the *Philosophical Transactions*, and the indications are that they fitted only uneasily into the Royal Society's work. Perhaps the most famous of this period are those of John Wallis and Christopher Wren on collision. Both papers are in Latin, like most other mathematical pieces, whereas English contributors on other topics usually wrote in their native language. The authority accruing to this material seems to have lain somewhat in doubt, since mathematical forms of argument and demonstration could not be fitted readily into the usual form of presentation. Accordingly, Wren's paper is prefaced by remarks, in English, calculated to make it more palatable. The theory is "verified by many Experiments, made by Himself and that other excellent Mathematician *M. Rook* before the Royal Society, as is attested by many Worthy Members of that Illustrious Body."[47] The departures in these mathematical papers from the usual format show that the Royal Society could not comfortably accommodate together the disparate traditions and criteria of experimental and mathematical researches.

Simon Schaffer's recent stress on the importance of the "matter of fact" both to Restoration science and to the ideological position of the Royal Society stems from his claim that the matter of fact posits "certain forms of social

---

46 Certain Fellows, in particular John Wilkins, advocated the development of a "plain style" of language in which to couch natural philosophy, and Wilkins, as is well known, attempted to develop a philosophical language. The welcome accorded the latter by the Royal Society was, however, less than resounding, and Hans Aarsleff suggests that this was due to its overly rationalist, essentialist implications. Aarsleff, "John Wilkins," DSB, vol. XIV, 361–381, which contains a wealth of relevant references, and M.M. Slaughter, *Universal Languages and Scientific Taxonomy in the Seventeenth Century* (Cambridge: Cambridge Univ. Press, 1982).

47 *Phil. Trans.* 3 (1669), 864 (Wallis), 867 (Wren), quoting 867. Perhaps this discomfort with mathematical material is connected with the distinction drawn by Thomas Kuhn in his "Mathematical versus Experimental Traditions in the History of the Physical Sciences," *Journal of Interdisciplinary History* 7 (1976), 1–31, reprinted in Kuhn, *The Essential Tension* (Chicago: Chicago Univ. Press, 1977), 31–65.

organization."[48] His remarks relate to the specificities of the English context, and are extremely valuable in that regard. My purpose in the present article, however, is to suggest themes constitutive of the Scientific Revolution as a whole, looking beyond the particular situation of the early Royal Society while using it as a case example. Whatever the concerns underlying the natural philosophy of the early Fellows, and the ideological concomitants of the Society's existence, the Royal Society was recognized throughout Europe as a body devoted to a form of natural inquiry, that had adherents in all parts of the continent. The Society's empiricism, especially that of Boyle, was closely paralleled by the approach in the 1650s of the Florentine Accademia del Cimento, a group operating in a very different social context from that of the English philosophers. The activities of the Royal Society were also consonant with those of the Académie des Sciences in this period, despite the widespread notion that the Académie was concerned preeminently with mathematical sciences.[49] The continental prestige of the Royal Society sprang from roots independent of its own domestic programs and concerns (although the latter are essential to a full understanding of science in Restoration England), and this article has been concerned with features of the natural philosophy of the Society which placed it in the wider context of all of Europe. The early Fellows certainly used the "matter of fact" to promote their own particular ideological ends, but they did not invent it; they deployed it. What Schaffer calls the "matter of fact" was, we have seen, constituted by a literary form which simultaneously redefined the role of experience in natural philosophy and provided a new basis of authority. That the Royal Society could itself exploit this redefinition of authority is suggested by Schaffer's remarks, but to understand its genesis requires a broader perspective.

The older forms of experience and authority were paradigmatically embodied in the genre of commentary. The new forms were embodied in a research report that provided details of discrete events. The change from one form to the other can be traced in the relative fortunes of these literary forms in natural philosophy. During the course of the seventeenth century, within the

---

48 Schaffer, "Making Certain," 140.

49 On the Accademia del Cimento, see W.E. Knowles Middleton, *The Experimenters. A Study of the Accademia del Cimento* (Baltimore: Johns Hopkins Univ. Press, 1971); and Richard Waller, *Essays of Natural Experiments* (London, 1684; facs. ed., New York: Johnson Reprint, 1964), a translation of the Accademia's *Saggi* into English under the auspices of the Royal Society. On the early Académie des Sciences, see Roger Hahn, *The Anatomy of a Scientific Institution. The Paris Academy of Sciences, 1666–1803* (Berkeley: Univ. California Press, 1971), Ch. I.

scholastic tradition itself, commentaries on Aristotle's natural philosophical works increasingly gave way to textbooks and compendia organized along thematic lines.[50] Although the substance of these works remained for the most part rooted in Aristotle and his commentators, the freedom of structure implied a loosening of the older forms of authority and, one may conjecture, encouraged the development of alternative techniques of justification. Mersenne, in many ways a transitional figure, moved from an apologetic defense of Aristotelian authority in the early 1620s to a position strikingly similar to that seen among the early Fellows of the Royal Society. In a letter to Jean Rey in 1632 he attempted to convince his correspondent that heavy bodies do not fall at a faster rate than light ones: "Truly, I am astonished at what you distrust of my experiment [*expérience*] of the equal speed of an iron bullet and a wooden bullet: for if several persons of quality who have seen and made the experiment with me were simply to swear [*signer*] solemnly to you, they would witness it to you authentically."[51] Authenticity here devolves from the witnessing of an event, not from the authorship of an authoritative text.

Because the changes in literary form and the changes in the effective meanings of authority and experience are firmly integrated, it is difficult to assign logical priority. They may not be separable at all, but together may be seen as constituting a new structure of natural philosophical discourse and practice. The reasons for the appearance of this new structure are hard to discern, but the social uses of this structure in the English context may point the way to a broader understanding of the Scientific Revolution. The importance of skeptical arguments in the sixteenth and seventeenth centuries arose in large measure, as Richard Popkin and others have shown, from their value in religious disputes. From Catholic attacks on Protestant theology in the sixteenth century to the Latitudinarians' probabilistic approaches to knowledge in Restoration England, the ability to cast general doubt on one's opponents' claims to knowledge served as a powerful weapon.[52] But in order to maintain a

50 See Patricia Reif, "The Textbook Tradition in Natural Philosophy 1600–1650," *Journal of the History of Ideas* 30 (1969), 17–32. John Heilbron notes the change among Jesuit writers in the seventeenth century from writing commentaries to writing treatises in natural philosophy; see Heilbron, *Electricity in the 17th and 18th Centuries. A Study of Early Modern Physics* (Berkeley/Los Angeles: Univ. California Press, 1979), 108–110.

51 Mersenne, *Correspondance du P. Marin Mersenne, religieux minime*, vol. III, ed. Cornelis de Waard (Paris: Centre de la Recherche Scientifique, 1969), 274–275. On Mersenne see esp. Robert Lenoble, *Mersenne, ou la naissance du mécanisme* (Paris: J. Vrin, 1943).

52 See esp. Richard H. Popkin, *The History of Scepticism from Erasmus to Spinoza* (Berkeley/Los Angeles: Univ. California Press, 1979), Ch. I, on p. 14; see also Schaffer, "Making Certain."

front against cognitive anarchy, criteria of certainty, albeit reduced, were required. In the France of the 1620s one finds the "mitigated skepticism" of Mersenne and Gassendi; in the England of the 1660s and 1670s one finds a kind of probabilism which allowed "moral certainty" to properly accredited experiential facts.[53] The authority in which that "moral certainty" resided for the Royal Society's Fellows was conferred by a postscholastic literary form which also defined a new meaning for "experience." The Society's cooperative venture was itself based on the ability of this new literary form to define the level at which cooperation could occur.

53 See Popkin, *History*, Ch. 7; for the humanist origins of this doctrine see Peter Dear, "Marin Mersenne and the Probabilistic Roots of 'Mitigated Scepticism,'" *Journal of the History of Philosophy* 22 (1984), 173–205; and on the 1660s, Shapin, "Pump and Circumstance."

CHAPTER 2

Rhetoric in the Early Royal Society

Richard Nate

In the history of science the Royal Society is known as one of the earliest European institutions dedicated to a systematic investigation of nature. It marks the beginning not only of an organized scientific community, but also of a theoretical reflection on scientific discourse. Considering the fact that the Royal Society has become famous for its hostility to rhetoric, the title of this essay may require some explanation. Recent studies have demonstrated that, although the authors of the early Royal Society attacked the "flourishes of rhetoric" in their writings, they did not hesitate to make use of figurative language themselves.[1] In so far as they employed rhetorical means to reject the use of rhetoric, their strategy may be called anti-rhetorical.[2] When Thomas Sprat called for a scientific prose in his *History of the Royal Society* (1667), which should avoid all "swellings of style," and compared this to a return to the state of a "primitive purity" of language,[3] his alliterations revealed that he did not hesitate to use rhetorical means in order to get his message across. Sprat characterized the supposed non-metaphorical style of science with a number of metaphorical adjectives. Thus, the "plain style" was described by him as a "close, naked, natural way of speaking."[4] Ultimately, these adjectives can be

1 Cf. Michael Cahn, *Der Druck des Wissens. Geschichte und Medium der wissenschaftlichen Publikation* (Wiesbaden: Reichert, 1991) and also Geoff Bennington, "The Perfect Cheat. Locke and Empiricism's Rhetoric," in *The Figural and the Literal. Problems of Language in the History of Science and Philosophy, 1630–1800*, eds. Andrew E. Benjamin et al. (Manchester: Manchester University Press), 103–123.

2 Heinrich F. Plett, "Rhetorik der Renaissance. Renaissance der Rhetorik," in *Renaissance-Rhetorik/Renaissance Rhetoric*, ed. Plett (Berlin/New York: de Gruyter, 1993), 1–20, here: p. 12. Anti-rhetoric also has a long tradition in European history; its beginnings are to be found in Plato's criticism of the sophists, cf. Peter L. Oesterreich, *Philosophen als politische Lehrer. Beispiele öffentlichen Vernunftgebrauchs* (Darmstadt: Wissenschaftliche Buchgesellschaft, 1994), 21 ff.

3 Thomas Sprat, *The History of The Royal-Society of London for the Improving of Natural Knowledge* (London: J. Martyn, 1667), rpt., eds. Jackson I. Cope and Harold W. Jones (St. Louis, Mo.: Washington University Press/London: Routledge & Kegan Paul, 1958), 113.

4 Sprat, *History*, 113. For a close analysis of Sprat's adjectives see Werner Hüllen, *"Their Manner of Discourse". Nachdenken über Sprache im Umkreis der Royal Society* (Tübingen: Narr), 107 ff.

 | DOI 10.1163/9789004283701_004

traced to an even older metaphor, namely that of linguistic transparency (*perspicuitas*) which had formed an element of the *genus humile* since antiquity.[5]

To argue, however, that the authors of the early Royal Society were self-contradictory in their rhetoric against rhetoric would imply that they attempted to extinguish rhetoric from any kind of discourse whatsoever. This is certainly not true. In an article of 1985, Brian Vickers argued that the early scientists' attitude towards rhetoric was not as negative as Richard F. Jones had suggested in his pioneering and influential studies.[6] The Royal Society opposed the obscure language of alchemists, cabbalists and the English followers of Paracelsus, for instance, but this did not mean that they were suspicious of metaphorical language in general. When the proponents of the New Science refuted such theories together with their corresponding ways of linguistic expression, they did not criticize rhetorical or poetical language as such. Although they rejected metaphors and comparisons as substitutes for scientific demonstration, they considered them appropriate for other communicative purposes, e.g. persuasive, didactic and aesthetic ones.

The fact that the authors of the Royal Society mainly focused on the style of science may have led to the conclusion that they considered the non-rhetorical language of science as a model for all kinds of communication. However, statements in which a plurality of discourse varieties is acknowledged pervade the writings of popularisers of the New Science, like Thomas Sprat or Joseph Glanvill, and also of philosophers like Francis Bacon, who may be regarded as the Royal Society's *spiritus rector*, Thomas Hobbes, who was not a member of the Society, but a strong advocate of the prevailing mechanistic philosophy, and John Locke, who was a member of the institution. These statements may be taken as a proof that the rhetorical principle of *aptum* was not abandoned in seventeenth-century linguistic thought. In his discussion on the functions

5 On the metaphor of transparency see also Wolfgang G. Müller, "Die traditionelle Rhetorik und einige Stilkonzepte des 20. Jahrhunderts," in *Die Aktualität der Rhetorik*, ed. Heinrich F. Plett. (München: Fink, 1996), 160–175, here: p. 164 ff., and Heinz L. Kretzenbacher. "Wie durchsichtig ist die Sprache der Wissenschaften?" in *Linguistik der Wissenschaftssprache*, eds. Harald Weinrich and Heinz L. Kretzenbacher (Berlin/New York: de Gruyter, 1995), 15–39.

6 See Brian Vickers, "The Royal Society and English Prose Style. A Reassessment," in *Rhetoric and the Pursuit of Truth. Language Change in the Seventeenth and Eighteenth Centuries*, Papers Read at a Clark Library Seminar, 8 March 1980, eds. Brian Vickers and Nancy S. Struever (Los Angeles: University of California, William Andrews Clark Memorial Library, 1985), 3–76 and his criticism of Richard F. Jones et al., *The Seventeenth Century. Studies in the History of English Thought and Literature from Bacon to Pope* (Stanford, Cal.: Stanford University Press/London: Oxford University Press, 1969), 75 ff., 143 ff.

of rhetoric on civil occasions, Bacon put it very clearly: "...if a man should speak of the same thing to several persons, he should nevertheless use different words to each of them...."[7] According to Bacon the truth value of a statement and its suitability in certain communicative situations should not be confused with each other. In the following it will be shown that, although the early scientists favoured the use of a plain style and thus adhered to the ancient principle of *perspicuitas*, they still believed in the relative appropriateness of styles and thus adhered also to the ancient rhetorical idea of *aptum*.

1 Anti-rhetoric and Its Background

Seventeenth-century polemics against rhetoric can only be understood when considered in the light of a specific understanding of rhetoric which had become popular since the sixteenth century. In the wake of Petrus Ramus, the notion of rhetoric had changed dramatically in that *inventio* and *dispositio* had been transferred to dialectic, and rhetoric had become more or less equated with *elocutio* and *actio*.[8] The ancient system of the *quinque partes artis*, which had included the processes of finding arguments and arranging them in the sequence of discourse, had been replaced by the elocutionary system of tropes and figures. Accordingly when members of the Royal Society launched their attacks on rhetoric they did not think of the strategies of *inventio*, *dispositio* or *memoria*, but first of all of *elocutio*.

An identification of rhetoric with *elocutio* is already apparent in Bacon's classification of the sciences. Bacon's arts of "Discovering," "Judging" and "Retaining" may still bear traces of the rhetorical *inventio*, *dispositio* and *memoria*, but they are now presented as subdivisions of logic, i.e. they are no longer counted among the arts of discourse. The art which is concerned with linguistic matters is called "Art of Transmission" and includes both grammar and rhetoric. The latter is defined as "illustration of discourse"[9] and more or less equivalent to the former *elocutio*.

Compared to the traditional rhetorical system, Bacon's classification is similar in terms of structure, but different in terms of content. *Inventio*, *dispositio* and *memoria* are defined in a non-linguistic sense, and only *elocutio* remains as

7 Francis Bacon, *The Works of Francis Bacon*, eds. James Spedding, Robert L. Ellis, and Douglas D. Heath, 14 vols. (London: Longman, 1858–1874), rpt. Stuttgart-Bad Cannstatt: Frommann-Holzboog, 1961–1963, vol. 4, 457 f.

8 Plett, "Rhetorik der Renaissance. Renaissance der Rhetorik," 5.

9 Bacon, *Works*, vol. 4, 454.

a linguistic art. Bacon's "Art of Transmission," however, can also be interpreted as a modification of the former *actio*, because it includes aspects of prosody and gestural communication.[10] Seen from this perspective, the former *elocutio*, i.e. Bacon's "illustration of discourse," is integrated into a theory of *actio* that has been extended into a complex theory of communication.

In order to understand Bacon's attitude towards rhetoric it seems useful to consider also his definition of the "Art of Discovering." When Bacon distinguished between the "Invention of Speech and Arguments" on the one hand and the "Invention of Arts and Sciences" on the other,[11] he may, with respect to the former, still have thought of rhetorical *inventio*, but at the same time he pointed out that only the latter had a scientific value: "The invention of arguments is not properly an invention; for to invent is to discover that we know not, not to recover or resummon that which we already know."[12] Here science and language are clearly set apart from each other, and it may therefore be concluded that Bacon saw science as non-rhetorical.[13] As will be shown below, the members of the early Royal Society followed him in this respect.

The critical attitude of seventeenth-century writers towards rhetoric is well documented. It is discernible in important philosophical works, e.g. Francis Bacon's *The Advancement of Learning* (1605), Thomas Hobbes' *Leviathan* (1651) or John Locke's *Essay Concerning Human Understanding* (1690), and in less famous tracts like Samuel Parker's *Free and Impartial Censvre of the Platonick Philosophie* (1666) or Joseph Glanvill's *Plus ultra* (1668). If these texts are viewed against their historical background, it becomes obvious that they did not attack rhetoric as such, but criticized an attitude towards rhetoric that was typical of certain philosophical schools. Thus, a criticism of rhetoric could either refer to the ornate style of the Ciceronians or to the highly metaphorical style of the occultists and hermeticists. Both schools were blamed for ascribing more importance to *verba* than to *res*. According to Bacon's view, the admirers of Cicero overemphasized the formal aspects of language, whereas the hermeticists refused to accept a distinction between words and things at all. When Bacon described "three vanities in studies, whereby learning hath been most traduced," he referred not only to the "contentious learning" of scholasticism

10 Bacon, *Works*, vol. 4, 438 ff.

11 Bacon, *Works*, vol. 4, 407.

12 Bacon, *Works*, vol. 4, 421. Cf. on this point also Brian Vickers, "Bacon and Rhetoric," in *The Cambridge Companion to Bacon*, ed. Markku Peltonen (Cambridge: Cambridge University Press, 1996), 200–231, here: p. 226.

13 In this respect his views differ from recent attempts to develop a rhetorical theory of enquiry, cf. the articles in *The Rhetorical Turn. Invention and Persuasion in the Conduct of Inquiry*, ed. Herbert W. Simons (Chicago/London: University of Chicago Press, 1990).

which according to his opinion produced merely "cobwebs of learning," but also to the "delicate learning" of the Ciceronians and the "fantastical learning" of "Astrology, Natural Magic, and Alchemy."[14]

In the mid seventeenth century, Bacon's thoughts were seldom repeated with the same clarity, but they still formed the background of anti-rhetorical statements. In the writings of Samuel Parker, Thomas Sprat and Joseph Glanvill it is clear that anti-rhetoric was used as a means to distinguish one's own position from other types of discourse. Rather than formulating positive descriptions of the supposed language of science, authors often defined it negatively by delineating it from rhetoric and poetry. What Sprat described as a "return to the primitive purity, and shortness, when men deliver'd so many *things*, almost in an equal number of *words*,"[15] implied a refutation of the Erasmian ideal of *copia rerum ac verborum* as well as a criticism of the hermeticists whose theories were considered as mere products of the imagination and thus compared to the fictions of poetical discourse. This may explain why Sprat, in describing the "model of their [i.e. the Royal Society's] whole design," mentioned rhetoric together with fancy and "Fables" as things which were to be avoided:

> Their purpose is, in short, to make faithful *Records*, of all the Works of *Nature*, or *Art*, which can come within their reach.... And to accomplish this, they have indeavor'd, to separate the knowledge of *Nature*, from the colours of *Rhetorick*, the devices of *Fancy*, or the delightful deceit of *Fables*.[16]

In a similar vein, Samuel Parker attacked the Rosicrucians, whom he regarded as the seventeenth-century followers of Plato, for being both rhetorical and poetical. In so far as they relied on metaphors and analogies as a means of philosophical proof, they could be accused of confusing facts with fictions:

> I can easily allow their Discourses the Title of *Philosophical Romances*... and 'tis with this estimate I would have them read: But when they pretend to be Natures Secretaries, & to understand all her Intrigues...and

14 Bacon, *Works*, vol. 3, 282 ff. For Bacon's relation to Ciceronianism see Judith Rice Henderson, "'Vain Affectations'. Bacon on Ciceronianism in the *Advancement of Learning*," *English Literary Renaissance* 25 (1995), 209–234, for his criticism on occultism see Brian Vickers, "Analogy vs. Identity. The Rejection of Occult Symbolism, 1580–1680," in *Occult and Scientific Mentalities in the Renaissance*, ed. Vickers (Cambridge: Cambridge University Press, 1984), 95–163, here: p. 133 ff.

15 Sprat, *History*, 113.

16 Sprat, *History*, 61 ff.

> yet put us off with nothing but rampant Metaphors, and Pompous Allegories, and other splendid but empty Schemes of speech, I must crave leave to account them (to say no worse) Poets & Romancers, true Philosophie is too sober to descend to these wildnesses of Imagination, and too Rational to be cheated by them.... Her Game is things not words.[17]

The rejection of rhetoric often had moral connotations. When Thomas Hobbes described metaphors as a kind of deceit, because they involved the use of words "in other sense than that they are ordained for," he defined them as a sort of lie.[18] According to the same principle Locke did not only denounce the use of figurative language, but also challenged the status of rhetoric as an art. Compared to the humanist ideal of *homo rhetoricus*, Locke's statements reveal a strong hostility to rhetoric:

> ...all the art of rhetoric, besides order and clearness, all the artificial and figurative application of words eloquence hath invented, are for nothing else but to insinuate wrong ideas, move the passions, and thereby mislead the judgment, and so indeed are perfect cheats.... I cannot but observe how little the preservation and improvement of truth and knowledge is the care and concern of mankind; since the arts of fallacy are endowed and preferred. It is evident how much men love to deceive and be deceived, since rhetoric, that powerful instrument of error and deceit, has its established professors, is publicly taught, and has always been had in great reputation....[19]

Parker was even more radical when he indulged in a polemics against the sect of "Platonick Philosophers." Parker's understanding of Platonism was deliberately vague; he used the term as a label for Paracelsians, Rosicrucians and religious fanatics. In this respect Parker's "Platonism" is comparable to what contemporaries would label as "Enthusiasm," a term which carried philosophical as well as religious meanings and which was used pejoratively.[20]

---

17 Samuel Parker, *A Free and Impartial Censvre of the Platonick Philosophie* (Oxford: Richard Davis, 1666), rpt. (New York: AMS Press, 1985), 73 ff. An apology of the Royal Society can be found on p. 45 in the same text.

18 Thomas Hobbes, *The English Works of Thomas Hobbes*, ed. William Molesworth, 13 vols. (London: Bohn, 1839–1845), rpt. Aalen: Scientia, 1966, vol. 3, 20.

19 John Locke, *The Works of John Locke*, 10 vols. (London: Thomas Tegg etc., 1823), rpt. Aalen: Scientia, 1963, vol. 2, 288 ff.

20 See Michael Heyd, "The Reaction to Enthusiasm in the Seventeenth Century. Towards an Integrative Approach," *Journal of Modern History* 53 (1981), 258–280.

Enthusiasts were criticized for mistaking their imaginations for revelations and metaphors and analogies for statements of truth. Both of these qualities made them also morally suspect. Thus, Parker did not hesitate to defame his philosophical opponents by accusing them for being lascivious and a threat to the soberness of reason:

> All those Theories in Philosophie which are expressed only in metaphorical Termes, are not real Truths, but the meer Products of Imagination.... Thus their [i.e. the Rosicrucians'] wanton & luxuriant fancies climbing up into the Bed of Reason, do not only defile it by unchast and illegitimate Embraces, but instead of real conceptions and notices of Things, impregnate the mind with nothing but Ayerie and Subventaneous Phantasmes.[21]

Parker's ideals of rationality and a pure descriptive language are set out here in highly rhetorical terms, and it can easily be recognized that the author furnished his criticism of rhetoric with those very analogies which he otherwise rejected as inappropriate in a philosophical argument.

Anti-rhetorical statements, like the ones cited here, have been taken as evidence of a self-contradiction inherent in the discourse of early modern science.[22] But, even if there is a tension in the texts as to what they say and how they say it, their supposed self-contradictory character remains as yet unproven. In the following, it will be shown that despite their anti-rhetorical attitude the authors of the Royal Society were certainly aware of the rhetorical character of their own writings. However, they regarded them as persuasive, apologetical or didactic rather than scientific in a strict sense. It follows that the Royal Society's attitude towards language was not as naïve as has sometimes been suggested.

2 The Plain Style and the Style of Science

Seventeenth-century concepts of a plain style have been dealt with in a number of studies,[23] therefore it may suffice here to mention only some of its

21 Parker, *Free and Impartial Censvre*, 75 ff.

22 See for instance Paul de Man's reading of Locke, "The Epistemology of Metaphor," *Critical Inquiry* 5 (1978), 12–30.

23 Cf. Jones, *The Seventeenth Century*, 75 ff.; Vickers, "The Royal Society"; Hüllen, *Their Manner of Discourse*, 98 ff., and Werner Hüllen, "Der komplexe Hintergrund des einfachen Stils. Zur theoretischen und stilistischen Begründung der modernen

characteristics. The fact that seventeenth-century authors demanded a plain style in scientific matters as well as in pulpit oratory indicates that the institutionalisation of the New Science cannot be isolated from the political context of the Restoration. A comparison between Sprat's chapter on the Royal Society's "Manner of Discourse" and Joseph Glanvill's *Essay Concerning Preaching* (1678) or the stylistic ideas embedded in his semi-fictional *Anti-fanatical Religion, and Free Philosophy* (1676) reveals close relations between the scientific and religious aspects of the rhetorical debate.[24] While Sprat tried to distinguish the scientists of the Royal Society from the Rosicrucians and hermeticists, Glanvill contrasted the sober minded Anglicans with religious fanatics who mistook their personal imaginations for divine revelations.[25] In the first two decades following the Restoration the political implications of such views are obvious. The Royal Society aligned itself with the Stuart king, from whom it had received the Royal Charter, no less than with the Church of England, whose official head was the king.[26]

In both scientific and religious matters the plain style was contrasted with obscurity. It thus conformed to the classical idea of *perspicuitas* which had already been opposed to *obscuritas* in ancient rhetoric.[27] Perspicuity may be understood in two respects. On the one hand it points to the fact that a writer is willing to lay open his or her intentions to the reader and to let him or her partake in his thought processes. In this respect it may be understood in terms of truthfulness and reliability. On the other hand it embodies the idea that

---

Wissenschaftssprache durch Francis Bacon und die Royal Society," *Geschichte der Sprachtheorie. Studien zum Sprachbegriff der Neuzeit*, ed. Ulrich Hoinkes (Münster/Hamburg: Lit), 31–46.

24 Joseph Glanvill, "An Essay Concerning Preaching," in *Critical Essays of the Seventeenth Century*, ed. Joel E. Spingarn, 3 vols. (Bloomington/London: Indiana University Press, 1957), vol. 2, 273–277; and Joseph Glanvill, "Anti-fanatical Religion, and Free Philosophy. In a Continuation of the New Atlantis," in *Essays on Several Important Subjects in Philosophy and Religion* (London: John Baker/Henry Mortlock, 1676), rpt. Bernhard Fabian (ed.) (Hildesheim: Olms, 1979), 41 ff.

25 Cf. Sprat, *History*, 37 ff.; Glanvill, "Anti-fanatical Religion," 29.

26 On the close relationship between an Anglican Latitudinarianism and the New Science cf. Vickers, "The Royal Society," 61 ff. This does not imply, however, that there were no Puritan members in the Society, cf. Charles Webster, *The Great Instauration. Science, Medicine and Reform 1626–1660* (London: Duckworth, 1975), 88 for a detailed account in this respect.

27 Cf. S. Michael Halloran and Merrill D. Whitburn. "Ciceronian Rhetoric and the Rise of Science. The Plain Style Reconsidered," in *The Rhetorical Tradition and Modern Writing*, ed. James J. Murphy (New York: Modern Language Association of America), 58–72.

there should not exist any linguistic barrier between the categories of the mind and the nature of things. Considered in modern terms, the first meaning relates to the interpersonal or pragmatic function of language, the second relates to its referential function.[28] The concept of the plain style included both aspects. Scientific discourse was considered as a moral enterprise,[29] and at the same time as a means that would lead directly to the nature of things.

Judging from the perspective of a broad understanding of rhetoric, it has to be acknowledged that the proponents of a plain style did not advocate a non-rhetorical language, but favoured the *genus humile* which was marked by its linguistic simplicity and the absence of figures and tropes. Considered in this way, the plain-style idea rested on a paradox which has been described as the "style of non-style."[30] Insofar as the plain style emerged as an attempt to hide the linguistic nature of scientific communication from the reader and make him believe that he could, as it were, penetrate directly through language to the world of things, it may be subsumed under the rhetorical strategy of *dissimulatio*.[31]

A practical consequence of the plain-style idea, which its supporters did certainly not foresee, was that in the attempt to be perspicuous, authors tended to comment exhaustively on their own way of writing. Thus, their texts often assumed a self-reflexive character. As Michael McKeon observes: "The fundamental trope of this antirhetorical style is the self-reflexive insistence on its own documentary candor as well as on the historicity of the narrative it transparently mediates."[32] By using this "fundamental trope," an author made clear that he considered his text scientific and that he himself wanted to be regarded as a member of the scientific community. At a time when the linguistic and stylistic norms of scientific discourse were not yet fully established,

---

28 When Herbert P. Grice regards perspicuity as a basic principle of everyday linguistic knowledge, this shows how firmly the idea is established in Western thought, Grice, "Logic and Conversation," in *Syntax and Semantics*, eds. P. Cole and J.L. Morgan (New York: Academic Press, 1975), vol. 3, 41–58. Grice does not present perspicuity as a principle which is absolute, but as one which is normally presupposed by communicants.

29 See Hüllen, *"Their Manner of Discourse,"* 98 ff.

30 Cf. Joseph Gusfield, "The Literary Rhetoric of Science. Comedy and Pathos in Drinking Driver Research," *American Sociological Review* 41 (1976), 16–34, here: 17.

31 Kretzenbacher describes this as "Stilistik der Sachlichkeit" and compares its function to that of a magic hood, "Wie durchsichtig ist die Sprache der Wissenschaften?" 18, 26. For a definition of *dissimulatio* see Heinrich F. Plett, *Einführung in die rhetorische Textanalyse* (Hamburg: Buske, 1991), 96 ff.

32 Michael McKeon, *The Origins of the English Novel 1600–1740* (Baltimore/London: Johns Hopkins University Press, 1987), 105.

meta-communicative devices—e.g. the promise to state nothing but the truth and to avoid obscurities of any kind—were employed to testify the scientific nature of a certain text. This could also lead to problems. Thus, self-reflexivity stood in clear contrast to the principle of linguistic economy—the rhetorical principle of *brevitas*—which was also counted as an ideal. In extreme cases, the ideal of perspicuity could result in a hyper-transparency which a few decades later would become a target of parody and satire. In *A Tale of a Tub* (1704) and in *Gulliver's Travels* (1726), Jonathan Swift's narrators mocked the moderns not only by what they said, but also by how they said it. The tale teller's "Inclination to expatiate upon the Beauty of my own Productions" and Gulliver's lengthy confessions of having related "plain Matter of Fact in the simplest Manner and Style"[33] may be regarded as parodies of a scientific prose that in extreme cases was hyper-transparent rather than perspicuous.

If the ideal of perspicuity has formed a basic element of scientific discourse since the seventeenth century, this does not mean that the language of science is not based on any stylistic norms. Investigations of scientific papers have pointed to the regularities in the structuring of scientific arguments and their textual presentation.[34] Scientific texts may be regarded as rhetorical in two respects. They normally rest on deliberate decisions as regards the choice of terminology and the arrangement of arguments, and they are persuasive in that they are written with the aim of convincing other members of the scientific community of the correctness of the ideas presented. If rhetoric is ubiquitous, as has been claimed, then scientific discourse cannot be considered as non-rhetorical.[35] From this premise, however, it does not follow that there are no differences between the language of science and other language varieties at all.[36] Scientific discourse is characterized by certain restrictions which

33 Jonathan Swift, *A Tale of a Tub to Which Is Added the Battle of the Books and the Mechanical Operation of the Spirit*, eds. A.C. Guthkelch and D. Nichol Smith (Oxford: Clarendon, 1958), 132; Swift, *The Prose Writings of Jonathan Swift*, 16 vols., ed. Herbert Davis (Oxford: Blackwell, 1939–1974), vol. 11, 291.

34 Kretzenbacher, "Wie durchsichtig ist die Sprache der Wissenschaften?" Cf. also Harald Weinrich's discussion of Watson and Crick's 1953 paper on the DNA, Weinrich, "Wissenschaftssprache, Sprachkultur und die Einheit der Wissenschaft," in *Linguistik der Wissenschaftssprache* (Berlin/New York: de Gruyter, 1995), 155–174. On the historical development of the scientific paper as a text form see Charles Bazerman, *Shaping Written Knowledge. The Genre and Activity of the Experimental Article in Science* (Madison, Wisc.: University of Wisconsin Press, 1988).

35 Alan G. Gross, *The Rhetoric of Science* (Cambridge: Harvard University Press, 1990).

36 Richard Nate, "Rhetorik und der Diskurs der Naturwissenschaften," in *Die Aktualität der Rhetorik*, ed. Heinrich F. Plett (München: Fink, 1996), 102–119.

distinguish it from other linguistic varieties.[37] The highly conventionalized form of scientific presentation, for instance, clearly limits the use of individual rhetorical strategies. As will be shown, the members of the Royal Society were aware of such differences.

3 Rhetorical Practice: Apologies, Panegyrics and Fictions

A closer look at the sources reveals that the textual practice employed by the authors of the early Royal Society was multi-shaped rather than one-dimensional. In this respect it may be useful to remember that the Royal Society did not only aim at an improvement of "experimental philosophy," but projected also an "English Academy." This was envisaged as an English equivalent to the *Académie française* that had been founded in 1635.[38] Sprat described the proposed academy as a "fixt, and *Impartial Court* of *Eloquence*; according to whose Censure, all Books, or Authors should either stand or fall."[39] It is obvious that the Royal Society's first historian did not speak of scientific discourse in this context. On the contrary, his proposal provides evidence that the Royal Society's criticism of rhetoric was not applied to every type of discourse. A few decades later, the idea of a "Court of Eloquence" would even be taken up by an author who has been known for being a strong critic of the Royal Society. In his *Proposal for Improving and Ascertaining the English Tongue* (1711), Jonathan Swift suggested the standardisation of the English language in a way that closely resembled Sprat's earlier views.[40]

It is evident that the early Royal Society developed a rhetorical practice of its own. Publications like Thomas Sprat's *History* and Joseph Glanvill's *Plus ultra* were clearly apologetical and written with the aim of persuading a wider public of the advantages that the scientific enterprise would bear. In the preface to his *History*, Sprat admitted that he had gone beyond the limits of the plain style, and he explained this by pointing to the persuasive character of his book. As Sprat observed in his preface, the qualities of "purity," "shortness" and

---

37 Kretzenbacher describes these restrictions as (1) "Ich-Tabu," (2) "Metapherntabu" and (3) "Erzähltabu," "Wie durchsichtig ist die Sprache der Wissenschaften?" 26 ff.

38 Hermann M. Flasdieck, *Der Gedanke einer englischen Sprachakademie in Vergangenheit und Gegenwart* (Jena: Frommansche Buchhandlung, 1928).

39 Sprat, *History*, 43.

40 Jonathan Swift, *A Proposal for Correcting, Improving and Ascertaining the English Tongue* (London: Benj. Tooke, 1712), rpt. Menston: Scolar Press, 1969.

"plainness" which were to characterize the scientists' "manner of discourse" could not be applied to his own way of writing:

> *The Style perhaps in which it* [i.e. the *History of the Royal Society*] *is written, is larger and more contentious than becomes that purity and shortness which are the chief beauties of Historical Writings*: *But the blame of this ought not so much to be laid upon me, as upon the Detractors of so noble an Institution* [i.e. the Royal Society]: *For their Objections and Cavils against it, did make it necessary for me to write of it, not altogether in the way of a* plain History, *but somtimes of an* Apology.[41]

Sprat's remarks reveal that the author did not regard his text as a part of scientific discourse, but understood it in terms of apology and persuasion. If considered in rhetorical terms, his text may be viewed as a mixture of three modes of representation. It belongs (1) to the *genus deliberativum* in its inclusion of a discussion of the principles of early modern science, (2) to the *genus judiciale* since it is constructed as a vindication of the New Science, and (3) to the *genus demonstrativum* as it praises the virtues of the proposed experimental philosophy and blames the vices of scholasticism and hermeticism. Rather than representing a typical example of scientific discourse, Sprat's *History* should be viewed as a rhetorical and, in parts, even a political text which has science as its subject.

A further example of the rhetorical practice of the Royal Society is Abraham Cowley's "Ode to the Royal Society" which was prefixed to Sprat's *History*. The fact that Cowley reflected on the Society's philosophical and linguistic programme in a panegyrical poem has been taken as evidence of a contradiction between content and form.[42] It would seem more appropriate, however, to describe it in terms of an aesthetic tension. Although Cowley's ode deals with the New Science and its discourse, it does not belong to the discourse of science. Furthermore, it is not a description of, but a panegyric on the New Science. As such it forms a part of the epideictic literature which saw a revival in the early years of the Restoration period.

41 Sprat, *History*, "An Advertisement to the Reader."

42 Cf. James A. Winn, *John Dryden and His World* (New Haven/London: Yale University Press, 1987), 132: "…the manner of his [i.e. Cowley's] poem in praise of Baconian clarity entirely contradicts its matter." See also Jürgen Kamm, *Der Diskurs des heroischen Dramas. Eine Untersuchung zur Ästhetik dialogischer Kommunikation in der englischen Restaurationszeit* (Trier: Wissenschaftlicher Verlag, 1996), 127.

Finally, there are those texts which deal with scientific matters in a fictional way. These start early in the century with Francis Bacon's *New Atlantis* (1627) and Francis Godwin's *The Man in the Moone* (1638)—texts which may be considered as early instances of science fiction.[43] In the middle of the century these were followed by two continuations of Bacon's *New Atlantis*, one of which was written by the anonymous R.H. (1660) and the other by Joseph Glanvill (1676).[44] The rhetorical character of these texts can hardly be disputed. They are written either to stimulate organized scientific research (Bacon), to promote and illustrate astronomical theories (Godwin), to call for governmental support (R.H.), or to separate scientific discourse from that of religious enthusiasts (Glanvill). Even if these texts do not represent scientific discourse in a strict sense, they may be regarded as examples of a rhetorical practice in the interests of early modern science.

The apologies of the New Science, the panegyrical poems, the scientific utopias and romances provide evidence that the early years of the Royal Society were marked by the presence of various text forms and discourses. It may even be argued that, despite the scientists' hostility to imagination and the products of fancy, the discourse of the New Science was instrumental in establishing a new poetical genre as soon as it was transferred to the realm of fiction.[45] As has often been observed, works like *Robinson Crusoe* and *Gulliver's Travels* are marked by the principle of minute observation of phenomena and the use of a descriptive style. In retrospect, these works indicate that the links between science and literature were much closer than the seventeenth-century polemics against rhetoric and the products of fancy might at first suggest.

4 Rhetorical Theory: Discourse Varieties

Although the idea of appropriateness did not play an important role in the stylistic theories of the seventeenth century, it was expressed by a number of authors. As has been said, the idea that one and the same "thing" could be expressed in a number of different ways had formed the basis of the ancient

43 Bacon, *Works*, vol. 3, 127 ff.; Francis Godwin, *The Man in the Moone. Or a Discovrse of a Voyage Thither by Domingo Gonsales The Speedy Messenger* (London: Ioshua Katon/Thomas Warren, 1638), rpt. Amsterdam: Theatrvm Orbis Terrarum/New York: Da Capo Press, 1972.

44 R.H. Esquire, *New Atlantis. Begun by the Lord Verulam, Viscount of St. Albans. And Continued by R.H. Esquire* (London: John Crooke, 1660); Glanvill, "Anti-fanatical Religion."

45 McKeon, *Origins of the English Novel.*

rhetorical idea of *aptum*. In the *Advancement of Learning*, Bacon wrote that, even if it was "some hindrance" in philosophical matters, an elaborate style was "a thing not hastily to be condemned." In the context of "civil occasions, … conference, counsel, persuasion, discourse, or the like" it could even be useful.[46] Accordingly, when discussing the "method of discourse" in his *De augmentis scientiarum* (1623), Bacon proposed a number of teaching styles which he defined with respect to specific audiences and communicative purposes.[47]

Hobbes listed four basic functions of language in his *Leviathan*. According to Hobbes, language may be used (1) "to register, what by cogitation, we find to be the cause of anything, present or past; and what we find things present or past may produce, or effect," (2) "to show to others that knowledge which we have attained, which is, to counsel and teach one another," (3) "to make known to others our wills and purposes" or (4) "to please and delight ourselves and others, by playing with our words, for pleasure or ornament."[48] The implication of these functions suggests four corresponding types of discourse: (1) the language of science, (2) the language of teaching, (3) private conversation, and (4) poetry. In a subsequent chapter, which included a discussion of fancy and judgment as two principal intellectual operations, Hobbes presented a further list of discourse varieties. Here, it was the relative proportion of fancy and judgment which constituted different types of discourse. Whereas poetical texts were characterized by a predominance of fancy, historical texts were ruled by judgment. Three other types listed by Hobbes may be compared to the classical rhetorical genres. Thus Hobbes' "orations of praise, and…invectives," "hortatives, and pleadings" and "demonstration" recall the *genus demonstrativum*, the *genus judiciale* and the *genus deliberativum*.[49]

In his *Essay Concerning Human Understanding* (1690), John Locke distinguished between a "philosophical discourse" on the one hand and a "civil discourse" on the other. According to Locke words may be distinguished:

46 Bacon, *Works*, vol. 3, 284.

47 Bacon, *Works*, vol. 4, 449 ff. Cf. also Hüllen, *"Their Manner of Discourse,"* 47 ff. for a discussion of these.

48 It is typical of the critical attitude in seventeenth-century linguistic thought that these uses of speech are contrasted with four kinds of abuse. The latter include (1) "inconsistency of the signification of…words," (2) "words used metaphorically" in order to "deceive others," (3) lying, and (4) words used "to grieve one another," Hobbes, *Works*, vol. 3, 20.

49 There are, of course, terminological differences. "Demonstration" is taken in a strictly philosophical sense by Hobbes; it is concerned with a "rigorous search for truth" and therefore not to be confused with the *genus demonstrativum*, i.e. the discourse of praise and blame.

> First, by their civil use, I mean such a communication of thoughts and ideas by words, as may serve for the upholding common conversation and commerce, about the ordinary affairs and conveniences of civil life, in the societies of men one amongst another. Secondly, by the philosophical use of words, I mean such an use of them as may serve to convey the precise notions of things, and to express, in general propositions, certain and undoubted truths, which the mind may rest upon, and be satisfied with, in its search after true knowledge. These two uses are very distinct; and a great deal less exactness will serve in the one than in the other.…[50]

Although there is little doubt that Locke was mostly interested in the philosophical use of words, he was aware that the existence of discourse varieties was something which could not be explained away. Even if Locke showed a strong anti-rhetorical attitude, the use of figurative language could not be condemned in any case:

> Since wit and fancy find easier entertainment in the world than dry truth and real knowledge, figurative speeches and allusion in language will hardly be admitted as an imperfection or abuse of it. I confess, in discourses where we seek rather pleasure and delight than information and improvement, such ornaments as are borrowed from them can scarce pass for faults.[51]

When Locke stated that the purpose of language was "first, to make known one man's thoughts or ideas to another; secondly, to do it with as much ease and quickness as possible; and, thirdly, thereby to convey the knowledge of things,"[52] he put human communication on the same level as an exchange of information and expressed a view of language that was mainly referential. It was on this theoretical basis that Locke also suggested his remedies of the "imperfection and abuse of words." More telling, however, is Locke's admission that an overall fixation of the meanings of words could never be expected. Thus it was only for "those who pretend seriously to search after or maintain truth" that the philosopher formulated his suggestions of a linguistic reform. Although Locke's linguistic ideal was a purely referential and standardized

50 Locke, *Works*, vol. 2, 251 ff.

51 Locke, *Works*, vol. 2, 288.

52 Locke, *Works*, vol. 2, 248.

language, he was realistic in his view of language use.[53] As has been shown already, he resembled other writers of the seventeenth century in this respect. Like the authors of the early Royal Society, Locke restricted his "window-pane-theory" of language[54] to the sphere of scientific discourse. As far as other language varieties were concerned, the principles of classical rhetoric were hardly called into question.

5 Conclusion

In the light of what has been said it seems reasonable to conclude that the members of the early Royal Society did not only use a number of rhetorical styles, but also developed a rhetorical theory to some extent. Although a plain style was demanded in the sphere of science that would direct the reader's attention away from language in order to lead him directly to "things," this was not intended to be a general linguistic norm. On the contrary, the scientists adhered to a communicative principle according to which different types of discourse were appropriate in different situations. When scientists and philosophers warned against the abuses of speech, they did not concern themselves with poetical language or with everyday communication, but with the language of science.

Nonetheless it is to be admitted that there was also a linguistic movement in the Royal Society in which the prevailing anti-rhetorical attitude *was* extended to other areas of culture. In the universal-language schemes, the scientific ideal of a non-rhetorical language indeed figured as a model for any kind of communication. In the preface to his *Essay Towards a Real Character and a Philosophical Language* (1668), John Wilkins envisaged a linguistic reform that would establish a transparent scientific discourse no less than social harmony and religious peace.[55] Such aims reveal the utopianism that characterized many of the projects of the early scientific community. Furthermore, the attempt to establish a language which according to Wilkins should not only be philosophical, i.e. scientific, but also universal, still bears traces of the dream of returning to a prelapsarian linguistic state.[56] As far as its original intentions are

53 A crucial example is Locke's discussion of the names of mixed modes which may even be regarded as an early instance of linguistic relativism, cf. Locke, *Works*, vol. 2, 195 ff.

54 Gusfield, "The Literary Rhetoric of Science."

55 John Wilkins, *An Essay Towards a Real Character and a Philosophical Language* (London: S. Gellibrand/J. Martyn, 1668), rpt. *The Epistle Dedicatory* (Menston: Scolar Press, 1968).

56 Cf. Richard Nate, *Natursprachenmodelle des 17. Jahrhunderts* (Münster: Nodus, 1993).

concerned, Wilkins' plan must be regarded as a failure—despite the fact that in other respects it represented an important step towards the establishment of scientific nomenclatures.[57] When Locke distanced himself from the language reformers of the early Royal Society, he revealed a scepticism that was closer to Francis Bacon and Thomas Hobbes. Locke confessed: "I am not so vain to think, that anyone can pretend to attempt the perfect reforming the languages of the world, no, not so much as of his own country, without rendering himself ridiculous."[58]

In the early eighteenth century the idea of creating a purely referential language for all imaginable purposes would serve merely as a source of literary satire. In *Gulliver's Travels* (1726), Jonathan Swift acquainted the reader with intelligent horses who professed to use language solely as a means of exchanging information—only to reveal the contradictions inherent in such an attempt. Although superficially the Houyhnhnms stand for the ideals of an enlightened age, Swift presents them as unenlightened and self-deceived. The intelligent horses deny that they are engaged in rhetorical matters because they view language solely as a means "to receive Information of Facts."[59] However, their "old Debate" about the nature of their savage counterparts,[60] the Yahoos, demonstrates that the horses suffer from a lack of self-awareness. It is precisely that lack of self-awareness for which the authors of the early Royal Society and generations of scientists following them often have been blamed. As has been shown above, however, the Houyhnhms' failure hardly applies to those early scientists who demanded perspicuity in scientific matters, but did not deny the rhetorical character of public discourse.

57 On Wilkins' linguistic model and its impact for the further development of scientific discourse see Hüllen, *Their Manner of Discourse*, 148 ff., 195 ff.; on its relationship to the development of scientific taxonomies, cf. Mary M. Slaughter, *Universal Languages and Scientific Taxonomy in the Seventeenth Century* (Cambridge: Cambridge University Press, 1982).

58 Locke, *Works*, vol. 2, 289 f.

59 Swift, *The Prose Writings*, vol. 11, 240.

60 Swift, *The Prose Writings*, vol. 11, 271.

CHAPTER 3

Language Reform in the Late Seventeenth Century

Ryan J. Stark

In the opening chapter [*Rhetoric, Science, and Magic in Seventeenth-Century England*], I focused upon the rise of the new plain style, emphasizing how modern experimentalists rejected both the tropes of magic and mystery and, in the opposite direction, the discourses of materialism and skepticism (e.g., sadducism, deism, and atheism). In this chapter, I concentrate on the culmination of rhetorical reform in early modern England: the Royal Society's plain language program. I show in particular how members of the organization and other like-minded writers used the new plain style in order to counteract bewitching idioms, which most intellectuals of the period saw as the primary linguistic cause of England's religious and social strife. My overarching argument is that a group of experimental philosophers—by creating a new understanding of style, and of language in general—brought about a paradigm shift in the English rhetorical tradition.

Of course, significant changes in the world require many different participants in many different capacities, and the seventeenth-century reformation in style is no exception. A small circle of scientific writers did not alter the linguistic cosmos by themselves. To appreciate how plainness emerged on a large scale requires at minimum a non-linear model of how ideas influence societies, not to mention an intuition about the workings of *Zeitgeists*. James Sutherland is a bit hyperbolic when he suggests that after 1660 English prose gets a fresh start.[1] He would have been more accurate to say that English philosophy of rhetoric begins anew around 1665, give or take several years, building upon Francis Bacon's key formulation of the split between words and things in *The Advancement of Learning*, and also drawing upon many other criticisms of enchanted rhetoric in the late Renaissance, including those offered by philosophers, orators, poets, and, most notably, preachers and theologians. That phrasing, however, is cumbersome. English rhetoric begins anew on a massive scale in the Restoration, when the occult Renaissance cosmos starts to collapse in mainstream intellectual circles. On the question of who deserves the most credit for conceptualizing this rhetorical reformation (though not the only credit), it is difficult, if not impossible, to identify any other group besides those working in the tradition of modern experimentalism, before and after the establishment of the Royal Society. By advancing plainness so

1 Sutherland, *On English Prose*, 67.

 | DOI 10.1163/9789004283701_005

effectively, experimental philosophers transformed the English rhetorical universe, laying groundwork for the rise of the Enlightenment linguistic sensibility broadly imagined.

Thomas Sprat's and John Locke's Arguments against Bewitching Rhetoric

Many writers advanced the cause of rhetorical plainness in the late seventeenth century, but the most famous argument arrives in *The History of the Royal Society* (1667), where Sprat claims that the Royal Society shapes a new "plain" style for a new epoch.[2] Sprat, to be sure, advertises the plain language program as much as he reports upon its widespread influence. And, more generally, he writes something other than disinterested history in his *History*, which should not be a surprise. His book is better seen as hopeful history, but with real momentum. In 1667, it is too early to say definitively that Sprat was correct about what he described. After all, the Royal Society was in its infancy, and the idea of stylistic plainness still met with considerable resistance from numerous magical and mystical philosophers, including occult writers such as Elias Ashmole and Thomas Vaughan, both of whom worked against plainness as members of the Royal Society.[3] Importantly, however, Sprat was in the process of becoming right as the idea of the new plain style slowly pervaded almost every aspect of Restoration scientific and rhetorical culture, after many decades of conflict between new philosophers and occultists of various sorts. While far from neutral in his vision of the world, Sprat captured the gist of a massive rhetorical shift, even a catastrophic shift, from the standpoint of occult mentalities.

Sprat attacks "tropes and figures" in what has become a well-known passage from the *History*, where he simultaneously advocates the new plain style.[4] After noting that rhetorical ornaments pose no threat in the "hands of Wise Men," but pose a serious threat in the hands of "the Wicked," Sprat comments upon contemporary rhetoric in England: "But now [tropes] are generally chang'd to worse uses: they make the fancy disgust the best things, if they come to sound, and unadorned: they are in open defiance against Reason [...and]

---

2 Sprat, *The History of the Royal Society* (London, 1667), 112.

3 On Ashmole, see Tobias Churton, *The Magus of Freemasonry. The Mysterious Life of Elias Ashmole. Scientist, Alchemist, and Founder of the Royal Society* (Rochester, VT: Inner Traditions, 2006).

4 Sprat, *The History of the Royal Society*, 112.

they give the mind a motion too changeable, and bewitching, to consist with right practice."[5] Sprat argues that certain applications of tropes must be avoided because they have a "bewitching" effect. In other words, he figures bewitchment—not elaborate syntax—as the antithesis of rhetorical plainness. He bolsters the point by describing non-plain uses of rhetoric as "evil," which resonates with his earlier term "wicked."[6] Sprat places the Royal Society's plain style movement within a larger theological struggle between good and evil, between the experiments of scientists and the experiments of preternatural charmers. Importantly, too, words such as "evil" and "bewitching" do not have light-hearted auras in Sprat's work. By "bewitching," he means bewitchment in the most obvious sense—that is—real witchcraft. A staunch Anglican in the same intellectual circle as John Wilkins and Joseph Glanvill, Sprat believes wholeheartedly in the efficacy of preternatural sorcery (e.g., the conjuring of demons, the casting of spells). Moreover, he clearly sees such diabolical activity as especially rampant in the Interregnum and early Restoration: rhetoric has "chang'd to worse uses." Demonic eloquence abounds, and rhetorical plainness becomes in this context the best linguistic hope against those nefarious idioms that infiltrate large segments of the English population, creating the lingering possibility of warfare and strife. The Royal Society's push for plainness is a direct push against demonry, Sprat contends. The real enemy of plainness in *The History* is the Enemy, a neglected but crucial aspect of the new philosophers' rhetorical reforms.

The emphasis upon bewitchment as the contrast to plainness goes a long way toward explaining what Robert E. Stillman presents as the puzzle of Sprat's argument: "The tension between the figural action of [Sprat's *History*] and its frequently expressed hostility to figuration needs to be explained, not explained away."[7] Brian Vickers identifies the apparent problem even more dramatically, suggesting that Sprat suffers from blatant hypocrisy: "Those who attack rhetoric [e.g., Sprat, Locke] continued to use it, and the pronouncements of an avant-garde elite are as little use then, as now, for recording the whole picture."[8] Thomas Conley makes a similar observation: "Nor can we overlook the fact that even those who joined the Society's resolution [to write plainly] continued, in many cases, the same sort of prose they had condemned."[9]

5 *Ibid.*, 111.

6 *Ibid.*

7 Stillman, *The New Philosophy and Universal Languages in Seventeenth-Century England*, 36.

8 Vickers, *In Defense of Rhetoric*, 201.

9 Thomas Conley, *Rhetoric in the European Tradition*, 169. For other readings that focus problematically upon Sprat's syntax, not his philosophy of rhetoric, see Richard Nate, "Rhetoric in

Sprat, however, is completely aware of his rhetoricity, but even more importantly, he is aware that it is a certain type of rhetoricity, a non-bewitching type, which is precisely the sort of rhetoric that he advocates in the passage. No tension whatsoever exists between Sprat's call for plainness and his ample use of tropes and figures, because his tropes and figures are plain. Sprat is not a hypocrite. The rhetoric he attacks is not the rhetoric he uses. The difference between plain and bewitching tropes is the key distinction here. Critics have not fully appreciated the extra level of rhetoricity that Sprat builds into his own analysis of rhetoric, which causes them to charge Sprat with some form of self-contradictory rhetorical behavior. Rather than contradicting himself, Sprat provides a spectacular rhetorical instance of his point. Sprat uses plain tropes to challenge bewitching tropes. He illustrates. In the process, he also warns against the true threat to the new plain style: witchcraft and its concomitant rhetorical practices of enchantment, hexing, and charming. Sprat makes a philosophical argument against demonic rhetoric in this renowned defense of plainness, not an argument against tropes, not an argument against elaborate syntax, and certainly not an argument against the rhetorical tradition.

---

the Early Royal Society," in *Rhetorica Movet*, eds. Peter Oesterreich and Thomas Sloane (Leiden: Brill, 1999), 215–233; Carey McIntosh, *The Evolution of English Prose, 1700–1800* (Cambridge: Cambridge University Press, 1998), 153; Werner Hüllen, "Sprat's Demand for a Plain Style Reconsidered," in *Papers in the History of Linguistics*, eds. Hans Aarsleff, Louis Kelly, and Hans-Josef Niederehe (Amsterdam: Benjamins, 1987), 247–262. For additional arguments that similarly misconstrue the nature of the Royal Society's plain language reforms, see Jeanne Fahnestock, *Rhetorical Figures in Science* (New York: Oxford University Press, 1999); John Rogers, *The Matter of Revolution. Science, Poetry and Politics in the Age of Milton* (Ithaca: Cornell University Press, 1996); Adam Potkay, *The Fate of Eloquence in the Age of Hume* (Ithaca: Cornell University Press, 1994); Robert Markley, *Fallen Languages. Crisis of Representation in Newtonian England, 1660–1740* (Ithaca: Cornell University Press, 1993); Debora Shuger, *Sacred Rhetoric. The Christian Grand Style in the Renaissance* (Princeton: Princeton University Press, 1988); Michael Srigley, "The Evolution of the Plain Style in the Seventeenth Century," *Studia Neophilologica* 60 (1988), 179–192; Peter Dear, "Rhetoric and Authority in the Early Royal Society," *Isis* 76 (1985), 145–161; Paul Arakelian, "The Myth of a Restoration Style Shift," *Eighteenth Century. Theory and Interpretation* 20 (1979), 227–245; Murray Cohen, *Sensible Words. Linguistic Practice in England, 1640–1785* (Baltimore: The Johns Hopkins University Press, 1977); Joan Webber, *Style and Self in Seventeenth-Century Prose* (Madison: University of Wisconsin Press, 1968); Frank Wilson, *Seventeenth-Century Prose. Five Lectures* (Berkeley: University of California Press, 1960); Wilbur Samuel Howell, *Logic and Rhetoric in England, 1500–1700* (Princeton: Princeton University Press, 1956); A.C. Howell, "*Res et verba.* Words and Things," *English Literary History* 13 (1946), 131–142; Hugh MacDonald, "Another Aspect of Seventeenth-Century Prose," *Review of English Studies* 19 (1943), 33–43; Joan Bennett, "An Aspect of the Evolution of Seventeenth-Century Prose," *Review of English Studies* 17 (1941), 281–297.

Sprat, in fact, insists that the writers of the Royal Society should "keep up the Ornaments of speaking," suggesting furthermore that "Rhetoric" is "a Weapon which may be as fully prouv'd by bad men, as good."[10] Sprat does the exact opposite of what he is often accused of doing. He calls upon members of the Royal Society to use tropes for the good, and to recover their usefulness, provided that they apply them in non-bewitching ways. While discussing Bacon's achievements later in the *History*, Sprat offers additional evidence that he does not abandon ornamentation. He praises Bacon's "Noble Labours in [natural] Philosophy," where Bacon is able "to express and adorn his thoughts" with "a vast Treasure" of "imaginings."[11] In other words, Sprat admires Bacon's facility with rhetoric, because Bacon uses tropes in a modern skeptical way, as mere adornment to decorate the substance of the idea. After pointing to several "purple passages" in the *History*, Richard Kroll makes a similar argument about Sprat's positive though qualified application of rhetorical figures: "What Sprat resists is not figuration as such, but its false application."[12] Yes! And, moreover, the false applications of tropes that most concern Sprat are the metaphors of witchcraft and occult philosophy.

Joseph Glanvill, another member of the Royal Society often criticized for using metaphors to denounce metaphors, also distinguishes between wicked and edifying applications of tropes, doing so by praising Sprat's eloquence in the *History*. Glanvill argues that the "style" of the "book hath all the properties" to "recommend," because it is

> not broken with ends of Latin, nor impertinent quotations; nor made harsh by hard words, or needless terms of Art; not rendered intricate by long Parentheses, nor gaudy by flaunting metaphors; not tedious by wide fetches of circumferences of Speech, nor dark by too much curtness of Expression; Tis not lose and unjointed, rugged and uneven; but as polite and fast as Marble; and briefly avoids all the notorious defects, and wants none of the proper ornaments of Language.[13]

Ornamentation does not trouble Glanvill, as is evidenced by this now famous metaphorical description of Sprat's plain style: "as polite and fast as Marble," a spectacular simile. The "flaunting metaphors" and "dark" expressions of Renaissance spell casters are a different issue entirely, however, not because they are metaphors as such, but instead because they carry with them the

10 Sprat, *History of the Royal Society*, 111.

11 *Ibid.*, 411.

12 Richard Kroll, *The Material World. Literate Culture in the Restoration and Early Eighteenth Century* (Baltimore: The Johns Hopkins University Press, 1990), 278.

13 Glanvill, *Plus ultra* (London, 1668), 84.

presuppositions of an occult sensibility. The other notorious defects to which Glanvill speaks presumably include metaphysical conceits ("wide circumferences"), mystical word plays, anagrams, cryptograms, and other rhetorical accoutrements of the charmed Renaissance cosmos. Glanvill replaces periaptic metaphors with the "proper ornaments of Language." He advocates a new philosophy of plainness, which has nothing to do with abandoning tropes. Rather, Glanvill applauds the proper application of tropes, all the while distancing English science from witchcraft and charming.

Sprat's and Glanvill's arguments shatter the widely held perception that members of the Royal Society are somehow anti-rhetorical.[14] They are not. Positive remarks about rhetoric by Sprat and Glanvill also dispel the idea that they promote some type of anti-eloquence. It is probably even misleading to suggest that the new philosophers reject the Erasmian idea of *copia rerum ac verborum*, though they certainly do not celebrate it. Sprat and Glanvill's main purpose is to advance a modern philosophical awareness of rhetoricity as mere adornment, not magical adornment. They are anti-occult. This holds for both scientific and non-scientific discourses. To this end, Sprat imagines a special committee of the Royal Society that might function along the lines of the Port-Royal Académie Française, a group established in Paris in 1638, and one organized to protect the French language. Sprat advises the Royal Society to formulate a similar committee, a "fixt and Impartial Court of Eloquence, according to whose Censure, all Books, or Authors, should either stand or fall."[15] Such a court of English eloquence was created a few years later, and John Dryden, among other notable intellectuals, served on it. But it is the scope of Sprat's remark which is especially significant, because the Royal Society appears as an arbitrator of all eloquence, not only scientific eloquence. The new rhetorical plainness functions as the instrument of learning's advancement in every philological context, "all Books, or Authors." The Royal Society's language reforms are imagined to be comprehensive, and this comprehensiveness again serves as strong evidence against attempts to limit the plain language reforms to narrowly scientific contexts.[16]

14 See Robinson, *The Establishment of English Prose*, 160; Shuger, *Sacred Rhetoric*, 164–165; Morris Croll, "Review of R.F. Jones's 'Science and English Prose Style,'" in *Seventeenth-Century Prose*, ed. Stanley Fish (New York: Oxford University Press, 1971), 91–92.

15 Sprat, *History of the Royal Society*, 43.

16 R.S. Crane tries to resolve Sprat's apparent stylistic hypocrisy by saying that the Society's plain style program targeted only scientific language, not all types of discourse. Crane, "Review of R.F. Jones' 'Science and English Prose Style,'" in *Seventeenth-Century Prose*, pp. 92–93. See also Conley, *Rhetoric in the European Tradition*, 169. Sprat's and Wilkins's own remarks about the scope of the plain language program, however, betray these arguments. When Sprat contends that esoteric styles "ought to be banish'd out of all civil

Following in Sprat's footsteps, Locke also critiques rhetorical bewitchment, all the while advancing the new plain philosophy of style. The often-quoted passage from his third book of the *Essay Concerning Human Understanding* (1690) provides a good starting point for analysis:

> If we would speak of things as they are, we must allow that all the art of rhetoric, besides order and clearness; all the artificial and figurative application of words eloquence hath invented, are for nothing else but to insinuate wrong ideas, move the passions, and thereby mislead the judgment; and so indeed are perfect cheats; and therefore, however laudable or allowable oratory may render them in harangues and popular addresses, they are certainly in all discourses that pretend to inform or instruct, wholly to be avoided; and where truth and knowledge are concerned, cannot but be thought a great fault, either of the language or the person that makes use of them... [Yet] eloquence, like the fair sex, has too prevailing beauties in it to suffer itself ever to be spoken against. And it is in vain to find faults with those arts of deceiving, where in men find pleasure to be deceived.[17]

Locke captures some of the key attitudes surrounding mainstream Restoration philosophy of language, including most markedly the distinction between words and reality, between ornamentation and "things as they are."[18] There is also undoubtedly an anxiety about feminine mystery in Locke, and one certainly discovers an aspect of that here. A cryptic reference to Eve's deceit of Adam appears in the subtext. If there is a moral to the last sentence, it is as straightforward as this: do not be so easily fooled by feminine guile. Negative feminine archetypes from classical literature should also come to mind. By feminizing the abuse of words, Locke recalls ancient warnings about sirens'

Societies, as a thing fatal to Peace and Good Manners," his critique extends beyond scientific discourse (111–113). Wilkins also attacks the "grand imposture of Phrases" that has "almost eaten our solid knowledge in all professions" (*Essay Toward a Real Character and a Philosophic Language* [London, 1668], 17–18). The plain style mandate extends to "all professions," not just scientific discourses.

17 Locke, *Essay Concerning Human Understanding*, ed. Peter Nidditch (Oxford: Clarendon, 1975), 508.

18 On Locke's philosophy and rhetoric, see Peter Walmsley, *Locke's Essay and the Rhetoric of Science* (Bucknell, PA: Bucknell University Press, 2003); Jules Law, *The Rhetoric of Empiricism. Language and Perception from Locke to I.A. Richards* (Ithaca: Cornell University Press, 1993); Hans Aarsleff, *From Locke to Saussure. Essays on the Study of Language and Intellectual History* (Minneapolis: University of Minnesota Press, 1982).

songs and Circe's enchantments, rhetorical acts that pretend and promise but do not deliver, or, worse yet, deliver and then double-cross.

Locke's attack upon seductive eloquence, however, goes much deeper than an anxiety about feminine guile. The main target of his argument is the Devil himself. Female figure casters and witches simply stand in for the "Prince of Darkness."[19] Readers discover this most important aspect of Locke's philosophy of rhetoric by leaving behind the often-quoted third book and examining his critique of enthusiastic eloquence in Chapter 19 of the fourth book, where he discusses more fruitfully the proper and improper uses of rhetoric. For Locke, enthusiastic orators who apply tropes in seductive ways "perswade" people by creating "false light," a rhetorical strategy inspired by the "Son of the Morning [i.e., the Devil]."[20] In brief, Locke worries about demonic eloquence. Rhetoric by itself does not concern him. Rather, fiendish sophistry does. And Locke illustrates the point with his own convincing rhetoric, his argument by analogy. He uses paradigms of true and false light to make the case against demonic persuasion, a form of eloquence that seduces listeners into misguided feelings of warmth and illumination. Locke summarily rejects such rhetoric: "To talk of any other light [besides the light of reason] in the Understanding is to put ourselves in the dark, or in the Power of the Prince of Darkness, and by our own consent, to give ourselves up to Delusion to believe a Lie."[21] Exactly as Sprat does, Locke targets rhetorical bewitchment. The "Prince of Darkness" encourages humanity to believe lies by dressing them in seductive tropes, a point that harkens back to the story of Eden, where the Devil tells the first lie and reveals himself as the Father of Lies. Locke also recalls in book 4 his earlier remark in book 3 about men who willingly believe the deceptions of seductive rhetoric, as they believe the deceptions of women. In both cases, he addresses inappropriately amorous eloquence, but in book 4 he exposes the Devil as the figure standing behind such improper "application[s] of words." Satan, in other words, attempts through sophistry to mislead humanity by subverting the God-given light of reason.

Akin to the harlot witch, the enthusiastic orator in book 4 represents one more variation upon the Lockean theme of demonic seduction and, ultimately, demonic possession. The Prince of Darkness leads individuals—and congregations of individuals—down the road of ruin, all the while pretending toward illumination, in what amounts to a prime example of demonic inversion: sweet seems bitter, and bitter seems sweet. Locke makes a theological argument

19 Locke, *Essay Concerning Human Understanding*, 703.

20 *Ibid.*

21 *Ibid.*

against bewitching rhetoric, using the specific example of the enthusiastic orator to show how the Devil attempts to invert divine impulses in language. Audiences who succumb to rhetorical seductions, however, often feel "as fully satisfied of the illumination, i.e., are as strongly perswaded" as those who "are enlightened by the spirit of God."[22] In one of the *Essay*'s unusually poetic passages, Locke argues that wicked orators literally spellbind audiences, causing them to plummet into the abyss to the exact degree that they feel illuminated by truth. This is demonic irony in the extreme, applicable to Faustus's situation, for example, or to Macbeth's possession, where the Thane of Cawdor feels as if he is becoming himself (i.e., Macbeth the king) to the exact degree that he is destroying himself. Locke's purpose is to explain and then admonish this form of rhetorical demonry, where diabolical forces mimic and simultaneously undercut the good.

Philosophers and historians of rhetoric should also hear in the background of Locke's passage an argument against the idioms of factional enthusiasm. For most new philosophers associated with the Royal Society, nonconformist rhetoric continued to represent a particularly widespread type of false illumination, putting England in grave danger, even several years after the Restoration. By the standards of Lockean philosophy of rhetoric, the nonconformist mystics and magicians claim to improve society to the exact degree that they destroy it. Nonconformist eloquence functions at the center of this dynamic, threatening to overpower the rational faculties of the self and of proper English society (i.e., Anglican government), for the bad moving of the will, to recall Bacon's rhetorical framework—which is essentially Locke's rhetorical framework.

Given that demons attempt to persuade humanity falsely, Locke also wonders aloud how anyone "shall…distinguish between the delusions of Satan, and the inspiration of the Holy Ghost."[23] Descartes had raised this same topic in the *Meditations on First Philosophy*, pondering how a man might discover that he was fooled into believing that he thought, assuming that a demon could accomplish such a powerful rhetorical ruse.[24] This is for Descartes the hypothetical counterpart to *cogito ergo sum*; that is, men and women might discover that they are fooled into thinking that they think, which also proves existence (one must exist in order to be fooled), but not the sort of existence that Descartes set out to prove. By "demon," too, Descartes points to an evil

22 *Ibid.*, 704.

23 *Ibid.*, 703–704.

24 Rene Descartes, "Second Meditation," in *Meditations on First Philosophy* (Cambridge: Cambridge University Press, 1996).

spirit in the most literal sense, as does Locke, who believes wholeheartedly in the reality of the Devil. These philosophers are informed by traditional Christian metaphysics, which includes the reality of Satan's diabolical schemes. Responding to the problem of differentiating between false and true illuminations, Descartes and Locke offer essentially the same philosophical-rhetorical answer. The God-given capacity of natural reason allows humanity to distinguish between true and false rhetoric—divine and demonic eloquence. Significantly, the new plain style (i.e., a modern, non-enchanted style) operates as the instrument of that rational faculty, and, moreover, operates as evidence of that rational faculty. Charmed eloquence functions contrarily as the rhetorical instrument of the Devil and his intermediaries: enthusiasts, witches, alchemists, and magicians. These demonic orators imitate the Divine, all the while trying to undercut God and all that is holy in humanity (i.e., the breath of God, the human idiom in its edifying manifestation). As readers might expect, Locke identifies in Satanic rhetoric a troubling combination of antithesis and irony. The concomitant rhetorical plainness that flows from reason, not the diabolical imagination, subsequently becomes the rhetorical signal that one is not under the Devil's spell. Female witches usually cast spells most effectively, which again is why Locke emphasizes the feminine aspect of guile in his famous passage in book 3. The guile in question, however, is demonic at heart, and it is this element that should be highlighted.

Locke thus advances the Royal Society's language reforms, rejecting rhetorical enchantment and favoring the new plain style. He takes as self-evident the assumption that demons work through language to mislead humanity. The cure for such insidious rhetorical behavior is the Royal Society's plain sensibility, or what Locke also refers to as the plain, historical method, an idea of language and reality grounded not upon the darksome world of witchery, but rather upon the physical-theological world of new science. Moreover, it should be clear by now that plainness is a philosophical category for the advancers of learning. Like Sprat, Locke continues to use a variety of rhetorical tropes and strategies, but he does so plainly, not purging tropes from discourse, but instead applying tropes in merely ornamental ways.

Plain Conceits in Modern Philosophical Poetry

As representative examples of poets touched by Baconian philosophy, Abraham Cowley and Samuel Butler play important roles in promulgating the new plain sensibility. Both poets critique Renaissance magic, all the while advancing a non-occult philosophy of rhetoric built upon scientific advancement and Anglican orthodoxy. They forward the momentum of rhetorical plainness in Restoration England, where the occult philosopher's charmed tropes and the witch's preternatural invocations give way to the experimentalist's mere ornaments.

As a preface to *The History of the Royal Society*, Sprat attaches Abraham Cowley's "To the Royal Society," a laudatory poem that recounts how new philosophers jettisoned the old world of enchantment.[25] Cowley's main purpose is to render in verse form the *Zeitgeist* of scientific experimentalism, which he accomplishes with considerable zeal. Concerning the rhetorical tradition in particular, Cowley aptly characterizes the Royal Society's language reforms, from the standpoint of the reformers themselves:

> So from all Modern Folies He [Sprat]
> Has vindicated Eloquence and Wit.
> His candid Stile like a clean Stream does slide,
> And his bright Fancy all the way
> Does like the Sun-shine in it play.
> l. 174–78

Sprat rescues "Eloquence and Wit" from "all Modern Folies," not only scientific ones. The Royal Society's language reforms extend beyond the confines of the laboratory, as the poet reports, touching every aspect of English rhetoric, which is essentially how members of the Royal Society saw their duty as language reformers. The point of the undertaking was to refine the English tongue broadly imagined. Even the faculty of fancy becomes in Cowley's verse a thoroughly rational one. The imagination is now regulated by the candid and clean style of the new experimentalism, which recalls—however faintly—Bacon's definition of rhetoric: "The application of reason to the imagination for the better moving of the will."[26] And neither is it a stretch to perceive in Cowley's epideictic poetry a premonition of that same apotheosis of science expressed in Edmund Halley's famous ode to Isaac Newton, attached to the first edition of Newton's *Principia* (1687).[27] The ode, originally composed in Latin, ends with a celebration of scientific advancement:

---

25 On Cowley and the new science, see Mary Elizabeth Green, "Abraham Cowley as Baconian Apostle," *Restoration* 10 (1986), 68–75; Robert Hinman, *Abraham Cowley's World of Order* (Cambridge: Harvard University Press, 1960), 92–134.

26 Bacon, *Advancement of Learning*, 64.

27 On poetry and new science, see Marjorie Hope Nicolson, *Newton Demands the Muse. Newton's Optics and the Eighteenth-Century Poets* (Princeton: Princeton University Press,

Then ye who now on heavenly nectar fare,
Come celebrate with me in song the name
Of Newton, to the Muses dear; for he
Unlocked the hidden treasures of Truth:
So richly through his mind had Phoebus cast
The radiance of his own divinity.
Nearer the gods no mortal may approach.
l.59–65

The new scientist, rather than the inspired poet, appears in this proto-Enlightenment verse as the greatest champion of radiant truth, which becomes increasingly mechanical in nature as the Age of Reason takes hold. The sublime orator, on the contrary, looks increasingly superstitious in this context, as the urbane Shaftesbury is quick to suggest: "Amid the several styles and manners of discourse or writing, the easiest attained and the earliest practiced was the miraculous, the pompous, or what we generally call the sublime."[28]

While Sprat functions in Cowley's ode as the Royal Society's vindicator of eloquence and wit, Bacon plays the role of revered precursor: "Bacon at last… arose" and expelled the "ghost[s]" and "monsters" that made "children and superstitious Men afraid" (l.43–57). Cowley credits Bacon with moving English civilization out of the wilderness of fable and error, banishing occult ideas to the island of low science and superstition. Bacon achieves this success, moreover, by wielding "the plain Magic of true Reasons Light," an odd expression at best, and, at worst, one that flirts with those Rosicrucians who desired to claim Bacon as one of their own (l.45). What does Cowley mean when he says that Bacon used plain magic? The phrase relates to the new plain philosophy of rhetoric, though cryptically. By the precepts of learning's advancement, "plain magic" means no magic at all in one's rhetoric, or magic in reverse, which is the experimentalists' rhetorical antidote to occult philosophy of language. Using the term "plain," Cowley signals a non-occult attitude toward the occult. Plain magic is the end of magic. It is the triumph of rationalism over sorcery. Through the unusual expression, Cowley reveals how Bacon counteracts rhetorical bewitchment (i.e., the new plainness), ingeniously foreshadowing Sprat's like-minded argument against linguistic bewitchment in the *History*.

1946). See also William Jones, *Rhetoric of Science. A Study of Scientific Ideas and Imagery in Eighteenth-Century English Poetry* (Berkeley: University of California Press, 1966).

28 Anthony Ashley Cooper, "Third Earl of Shaftesbury," in *Characteristics of Men, Manners, Opinions, and Times*, ed. Lawrence E. Klein (Cambridge: Cambridge University Press, 2000), 108.

"To the Royal Society" is Cowley's best-known defense of science and, concomitantly, of the new plain style, but he also takes up the issue of learning's advancement in his encomium to William Harvey, which should be read as a sister poem to the ode. Cowley portrays the empirical researcher Harvey chasing the scantily clad nymph Daphne into a tree, only to discover small fibers, presumably to his disappointment:

Coy Nature, (which remain'd, though aged grown,
A Beauteous virgin still, injoy'd by none,
Nor seen unveil'd by any one)
When *Harveys* violent passion she did see,
Began to tremble, and to flee,
Took Sanctuary like *Daphne* in a tree:
There *Daphnes* lover stop't, and thought it much
The very Leaves of her to touch,
But *Harvey* our *Apollo*, stopt not so,
Into the Bark, and root he after her did goe:
No smallest Fibres of a Plant,
For which the eiebeams Point doth sharpness want,
His passage after her withstood.
1. l.1–13

Harvey uncovers a world of natural wonderment (smallest fibers), but not a world of magic. The poet juxtaposes in this scene two concepts of nature, one empirical and the other enchanted, favoring the former at the expense of the latter. As Cowley figures the passage, Harvey dispels all things occult. He represents another Bacon archetype, an experimentalist who banishes chimeras of the charmed imagination, and, perhaps, banishes the active role of Spirit in nature, which is a far more ominous underplot in the work. The argument culminates in a passage where the poet contrasts Harvey's innovative methods with the old rituals of Renaissance occultism:

Methinks in Arts great Circle others stand
Lock't up together, Hand in Hand,
Everyone leads as he is lead,
The same bare path they tread,
And Dance like Faries a Fantastic round,
But neither change their motion, nor their ground:
Had Harvey to this road confin'd his wit,
His noble circle of the Blood, had been untroden yet.
4. l. 59–66

Using the image of dancing fairies, Cowley alludes to ceremonial magic and alchemy. "Fairies" dancing in a "Fantastic" circle would have struck his audience as a magical procedure of some kind, a reference to a playful Puck, perhaps, or maybe a reference to sorcerers celebrating a pagan holiday under the full moon. The tone of the passage seems more serious than that of a midsummer night's dream. It is closer to the tone of Plato's *Ion*, where maidens dance frenetically by the riverside. Or, maybe Shakespeare's *Macbeth* should come to mind, especially those scenes where witches undulate in a fantastic circle on the murky heath. However we characterize the "Fantastic round," Cowley clearly makes a sober point. The royal physician, the subject of the poem, departs from the arcane heuristic procedures of the enchanted Renaissance cosmos and consequently discovers the blood's circulation. That is, Harvey rejects occult methods. Using epideictic rhetoric, Cowley celebrates this advancement of learning, providing as well a significant lesson about the nature of rhetorical style. From a purely formal standpoint, Harvey's model of circulation and the old model of circulation marked by dancing fairies both involve circles. The shape of the circle in and of itself, however, is no indicator of superstition, just as—by analogy—the formal shape of a sentence is no indicator of superstition. In this case, form must be informed by a philosophy of *forma*. Harvey produces a new circle, a plain circle, for a modern world, challenging the magical connotations of old models of circulation, physical, and, presumably, social and metaphysical.

Harvey is often credited with reinventing the internal nature of human biology, but less is said about the external implications of his discovery. By grounding the impetus of circulation within the body itself, Harvey, in one swift blow, shatters Renaissance conceptions of physic, including magical and astrological thinking based upon macrocosm to microcosm correspondences, where objects like the stars and the planets are thought to influence the body's and society's agency through occult verisimilitudes. In the absence of these esoteric systems of correspondences and superlunary persuasions, new models of circulation, physiognomic and ontological (e.g., gravity), begin to emerge in seventeenth-century England. This is due in no small part to Harvey's pioneering concept of internal impetus. The ancient categories of superlunary and sublunary existence give way to a modern theory of the cosmos, a place where the origins of motion in the body, and the body politic, are no longer grounded upon occult relationships with the crystalline sphere. This makes possible different avenues of agency and causation, including different ideas about how motion and circulation work. In the context of the rhetorical tradition, for example, Harvey's discovery reshapes ideas about how language moves the heart, not with the help of the stars and the planets, as Paracelsian alchemists believed, but rather with the guidance of the mind's sphere and other spiritual-biological structures. Harvey's internalized universe of volition anticipates the rise of faculty psychology and the concomitant theory of rhetoric contained therein, culminating in the English rhetorical tradition in George Campbell's *The Philosophy of Rhetoric* (1776).

By praising Sprat and Harvey, Cowley participates in this scientific trajectory. He writes as a modern philosophical poet, dismissing Interregnum *furor poeticus* and embracing the Baconian advancement of learning, which is advancement in all areas of study, including poetics. Poets such as John Donne, Richard Crashaw, Henry Vaughan, and Andrew Marvell appear increasingly superstitious against the backdrop of the new plain language reforms. In particular, their metaphysical conceits have all of the trappings of Cornelius Agrippa's occult philosophy of verisimilitudes, or of Marsilio Ficino's enchanted system of correspondences. In fact, this commonplace association between the metaphysical poets and occult practitioners is what caused metaphysical poetry to fall so far and so fast in Restoration England. Cowley's reputation as a poet did not suffer, however, despite his ingenious habits (e.g., "plain magic"). Samuel Johnson, the sometime spokesman of Enlightenment values, even positions Cowley as the greatest poet of his generation.[29] Why did not Cowley experience the same fate as many other metaphysical poets, to whom he is often compared? The brief answer is that his philosophy of rhetoric protected him. By clearly differentiating his poetic project from the projects of mystically inclined poets like Donne and Crashaw, Cowley makes himself a forerunner of modern politeness, as opposed to a twilight figure such as Vaughan, a poet who neither achieved nor desired that Restoration air of rhetorical austerity. Put differently, a mystical feeling lingers in the language of mainstream metaphysical poets and preachers. The same cannot be said of Cowley.

And neither does Samuel Butler write in the metaphysical mode. Butler, the cantankerous satirical poet most famous for *Hudibras*, anticipated many of the Royal Society's anti-magical sentiments, and he encouraged them, setting the stage for critiques of occult rhetoric in Cowley's "To the Royal Society" and "Upon Dr. Harvey," not to mention other modern philosophical poetry by Dryden, Swift, and Pope. Never a member of the Royal Society, Butler nonetheless played a major role in the Restoration plain language reforms, because he so forcefully reprimanded the occult Renaissance cosmos. In 1662–1663, he published the first two parts of *Hudibras*, a hugely popular mock epic attacking magical rhetoric and bad politics, which for Butler were inextricably linked.

29 Samuel Johnson, "The Life of Cowley," in *The Lives of the English Poets* (London, 1781).

The staunch Anglican begins the poem with a lengthy reflection upon the rhetorical tendencies of the not-so-heroic hero of the piece, whom he quickly associates with the world of magic:

> For RHETORIC, he could not ope
> His mouth, but out there flew a trope;
> And when he happen'd to break off
> I' th' middle of his speech, or cough,
> H' had hard words, ready to show why,
> And tell what rules he did it by;
> Else, when with greatest art he spoke,
> You'd think he talk'd like other folk,
> For all a rhetorician's rules
> Teach nothing but to name his tools.
> His ordinary rate of speech
> In loftiness of sound was rich;
> A Babylonish dialect,
> Which learned pedants much affect.
> It was a parti-colour'd dress
> Of patch'd and pie-bald languages;
> 'Twas English cut on Greek and Latin,
> Like fustian heretofore on satin;
> It had an odd promiscuous tone,
> As if h' had talk'd three parts in one;
> Which made some think, when he did gabble,
> Th' had heard three labourers of Babel;
> Or CERBERUS himself pronounce
> A leash of languages at once.
> This he as volubly would vent
> As if his stock would ne'er be spent:
> And truly, to support that charge,
> He had supplies as vast and large;
> For he cou'd coin, or counterfeit
> New words, with little or no wit:
> Words so debas'd and hard, no stone
> Was hard enough to touch them on;
> And when with hasty noise he spoke 'em,
> The ignorant for current took'em;
> That had the orator, who once
> Did fill his mouth with pebble stones

When he harangu'd, but known his phrase
He would have us'd no other ways.
1.82–118

Butler compiles a shorthand list of commonplace arguments against occult rhetoric that had already been made by figures such as Bacon, Wilkins, and Méric Casaubon. As one who popularized modern experimental attacks upon the vocabularies of magic, however, Butler provides an especially good summary of two key and interrelated ideas: occult rhetoric as babble and occult rhetoric as forgery. Butler first connects enchanted discourses to various "piebald" phenomena. Not surprisingly, the Tower of Babel looms as the dominant archetype throughout the critique. For Butler, as for most advancers of learning, occult rhetoric appears as an unnatural hodgepodge of parts, where spell casters assemble tropes and phrases in bizarre arrangement schemes, producing not eloquence but disorganized prattle.[30] By Butler's precepts, magical rhetors create Cerberus-like discourses, incoherent canting in several directions, which is an apt Anglican characterization of nonconformist illuminist rhetoric.

Secondly, Butler accuses seventeenth-century occult writers of counterfeiting hard words in order to obfuscate the true emptiness of their philosophical systems. Magicians and astrologers appear throughout *Hudibras* as rhetors of false restoration, promising valuable rhetoric but delivering instead sound and fury signifying nothing. Using the phrase "hard words," Butler also and surprisingly recalls the inkhorn controversies of the sixteenth century.[31] Anxieties about inkhorn rhetoric as such had subsided by the time Butler was writing, due in part to the success of Cawdry's *Alphabetical List* of hard words in 1605, the first monolingual dictionary in English.[32] Nonetheless, a similar type of anxiety about continental vocabularies re-emerges during the English Interregnum, though on a smaller and more focused scale. In all times of national stress, debates about proper idioms and external influences upon those proper idioms intensify, and the English Interregnum and early Restoration are no exceptions. In the case of the 1640s–1660s, and from the vantage point of the Anglican Royalists, the esoteric vocabularies of astrology and magic contend for prominence with the rhetoric of the new English experimentalism. For the

30 On Butler and rhetorical controversy, see Alvin Snider, *Origins and Authority in Seventeenth-Century England. Bacon, Milton, Butler* (Toronto: University of Toronto Press, 1994), 204–243; Ian Jack, *Augustan Satire. Intention and Idiom in English Poetry, 1660–1750* (Oxford: Oxford University Press, 1952), 27, 65–74.

31 See Wilson, *Arte of Rhetorique* (London, 1660), 164–165.

32 Robert Cawdrey, *Alphabetical List* (London, 1605).

soon-to-be philosophers of the Royal Society, the threat to the English language (and to the English way of life) is not exactly inkhorn rhetoric as John Bale, John Cheke, or Thomas Wilson understood it, but rather the threat is a related type of discourse, that is, the rhetoric of continental *esoterica*. In Butler's context, rhetorical diabolism is best exemplified by the circulation of Rosicrucian magic, theurgy, chiromancy, Catholic-style mysticism, and other habits of thought that work against Anglican orthodoxy. These idioms of magic and mysticism are the primary rhetorical targets in *Hudibras*.

Butler's reference to Demosthenes also proves interesting for reasons that go beyond the rejection of occult heuristics. The image allows Butler to criticize pronunciation, an often-overlooked aspect of magical rhetoric. He uses Demosthenes satirically to comment upon how contemporary orators might achieve the sort of pronunciation demanded by Rosicrucian spell casting. The hard words of magic prove especially sensitive to articulation. Against the backdrop of the occult arts, pronunciation takes on special connotations that have long since vanished in modern speech theory. To pronounce words correctly activated magical efficacy. Contrarily, to mispronounce words risked magical inefficacy, or, worse yet, magical misfire. Pronunciation takes on a less significant role in a paradigm where words and sounds are merely dress for thought, not entelechial expressions of a rhetorical-cosmological harmony, or what the young Glanvill refers to in *The Vanity of Dogmatizing* as a magical pitch.[33] Mispronunciation may risk misunderstanding or embarrassment in modern rhetorical paradigms, as in the case of malapropisms, but it does not risk catastrophe: turning lead into something other than gold, knocking down walls inadvertently, transmogrifying audiences into beasts, corrupting sacraments, and so forth. In mystical and magical paradigms where sounds, pitches, and pronunciations participate in transforming the world, the requirements of oral rhetoric take on an additional layer of significance. In the occult Renaissance cosmos, mispronunciation becomes a mis-making of the cosmos. As Vickers rightly suggests about Renaissance magic, "to rearrange words is to rearrange reality."[34] Along the same lines, to rearrange sounds is to rearrange reality, which is why plays on sound hold such a provocative place in medieval and Renaissance eloquence. Butler understands this crucial role of paronomasia in magical rhetoric, and he relentlessly parodies it.[35]

---

33 Joseph Glanvill, *The Vanity of Dogmatizing*, 198.

34 Brian Vickers, "Analogy versus Identity," in *Occult and Scientific Mentalities in the Renaissance*, 106.

35 The reader may also and rightly detect a subtle critique of Catholic transubstantiation at work in Butler's reference to putting pebbles in one's mouth, or, concomitantly, putting

The decline of occult philosophy in the seventeenth century is undoubtedly a key factor in the diminishing interest in the power of oral discourse, which is a lesson implicit in Butler's critique. Perhaps the ruin of academic magic is even more significant to the topic of orality and literacy than the invention of the printing press. Walter Ong and most other historians of rhetoric point to the mass circulation of the printed word as the principal force in diminishing the significance of the spoken word in Renaissance and Enlightenment rhetoric, but the collapse of magic is probably even more substantial.[36] Nothing technological prevents a reader from reading the printed word aloud, but something philosophical does. The "curiously silent" world of the new science, as Ong describes it, is a world without spell casting, where reading aloud and reading silently is a matter of preference, not a matter of reality making.[37]

Butler's satire on pronunciation repudiates occult ideas about sound. That sound is an incidental rhetorical property for Butler becomes amusingly clear when he reveals his own willingness to change names in order to achieve a rhyme scheme.

> A squire he had, whose name was RALPH,
> That in th' adventure went his half:
> Though writers, for more stately tone,
> Do call him RALPHO; 'tis all one;
> And when we can with metre safe,
> We'll call him so; if not, plain RALPH.
>
> 1.457–462

In the alteration of the name, the poet neither gains nor loses anything essential. To rearrange words and sounds is not to rearrange reality in the modern rhetorical universe. Compare the passage above with Herbert's mystical exploration of rhyme at the conclusion of "Denial," one of the numinous poems in *The Temple* (1633):[38]

---

the hard words of continental magic and mysticism in one's mouth. "Transubstantiation" is one of those "hard words" to which Butler refers. The sacrament appears indirectly in Butler's satire as a mere stone, or, in a more disturbing light, as mere loam.

36 Walter Ong, *Orality and Literacy. The Technologizing of the Word* (London: Methuen, 1982).

37 Ong, *Ramus. Method, and the Decay of Dialogue* (Cambridge: Harvard University Press, 1958), 118.

38 George Herbert, *The Temple* (London, 1633).

Therefore my soul lay out of sight,
  Untuned, unstrung:
My feeble spirit, unable to look right,
  Like a nipped blossom, hung
    Discontented.

O cheer and tune my heartless breast,
  Defer no time;
That so thy favors granting my request,
  They and my mind may chime,
    And mend my rime.
    l. 16–25

Herbert uses rhyme to suggest an occult sympathy with God's mystical order. After the first four stanzas of the poem conclude with unrhymed frustration, the poet finds his way back to heavenly music in the last stanza, by the grace of God, who "mend[s]" the poet's rhyme and heart. Herbert's rhymes are entelechial and reverberant, reinforcing the occult rhetorical principle of *similia similibus* (like produces like), which is one of the cornerstones of cosmic harmony in charmed philosophies of rhetoric. Butler's rhymes have no such sublime concord. They are purely utilitarian.

Butler accomplishes in satirical form what Cowley accomplishes straightforwardly—a repudiation of magical rhetoric. Of course, the satirical sting of a mock epic loses its force if it goes on for too long. Pope apprehends this point more clearly than Butler. *The Rape of the Lock* is a masterpiece. *Hudibras* is not, especially by the time Butler adds a third canto nearly a decade and a half after the original publication of the poem, in what amounts to a less than seamless sequel. *Hudibras* becomes its own kind of Cerberus, ironically. Nonetheless, on the topic of early Restoration rhetorical reform, Butler plays a far more important role than is usually acknowledged. He popularizes anti-magical attitudes toward language, calling into doubt the magical rhetoric of spell casters, which sets the stage for the Royal Society's new plain style. Through stable irony, Butler forwards the anti-occult sensibility of the Restoration language reformers, and so he participates firsthand in the advancement of modernity.

Preaching, New Science, and Plain Language

R.F. Jones argues that preachers sympathetic to experimental philosophy played the greatest part in circulating the new plain style in late seventeenth-century England.[39] This claim has been challenged recently, though it hardly seems controversial, if the idea is that Anglican preachers had a longer reach in the Restoration than other groups, which is Jones's contention.[40] Bishops shaped the agenda in many ways, including rhetorically. I agree with the spirit of Jones's thesis, except on the most vital point: he mistakenly speaks about a syntactical revolution, whereas I am arguing for a philosophical-rhetorical revolution. Such sweeping claims are difficult to prove, but no one had a more extensive influence than preachers. Of most interest are those eloquent bishops serving as members of the Royal Society. In particular, John Wilkins, John Tillotson, and Gilbert Burnet played significant roles in theorizing and circulating the new plain sensibility. Moreover, and most importantly for my thesis, plainness appears as a decidedly philosophical category in their writings, rather than a narrowly syntactical one. Theologians sympathetic to modern philosophy (mostly latitudinarian in temperament) advanced plainness in order to counteract occult philosophies of rhetoric.[41] Modern preachers expressed the same overriding concern that every other advancer of learning expressed, and this is an anxiety about the charmed Renaissance cosmos and the esoteric rhetorical activities contained therein.

In *A Discourse Concerning the Gift of Prayer* (1653), Wilkins argues for "plain English" in religious rhetoric, and in all writing and speaking, for that matter. Simultaneously, he argues against stylistic "Affectation," by which he means

39 Richard F. Jones, "The Attack on Pulpit Eloquence in the Restoration. An Episode in the Development of the Neo-classical Standard for Prose," in *The Seventeenth Century. Studies in the History of English Thought and Literature from Bacon to Pope* (Stanford: Stanford University Press, 1965), 111–142. See also Jones, "The Rhetoric of Science in England of the Mid-Seventeenth Century," in *Restoration and Eighteenth-Century Literature. Essays in Honor of Alan Dugald McKillop*, ed. Carroll Camden (Chicago: University of Chicago Press, 1963), 5–24.

40 Robert Markley, *Fallen Languages*, 2.

41 For additional arguments against nefarious eloquence by preachers sympathetic to the new experimentalism, see Isaac Barrow, *Several Sermons a against Evil-Speaking* (London, 1678); Richard Allestree, *The Government of the Tongue* (London, 1667). On latitudinarianism, religion, and science in seventeenth-century England, see John Spurr, "Rational Religion in Restoration England," *Journal of the History of Ideas* 49 (1988), 563–585; Lotte Mulligan, "Anglicanism, Latitudinarianism, and Science in 17th-Century England," *Annals of Science* 30 (1973), 213–219; Barbara Shapiro, "Latitudinarianism and Science in Seventeenth-Century England," *Past and Present* 40 (1968), 16–41; R.S. Westfall, *Science and Religion in Seventeenth-Century England* (New Haven: Yale University Press, 1958). See also Brian Young, *Religion and Enlightenment in Eighteenth-Century England* (Oxford: Oxford University Press, 1998); Martin Griffin, *Latitudinarianism in the Seventeenth-Century Church of England* (Leiden: Brill Academic Publishers, 1992).

> either too much neatness and elegance, or else a mystical kind of phrase, not to be found either in Scripture, or any sober Writer (though much in fashion amongst some men in these times), which, it may be, sounds well to vulgar ears; but being reduced into plain English, will appear to be wholly empty, and to signifie nothing, or else to be full of vain repetitions.[42]

The first type of affectation needs little explanation. Wilkins has in mind a courtly style that appears too easily as artifice. This is what Morris Croll refers to as the neo-Ciceronian style of the sixteenth century, and what Ian Robinson describes more accurately as the pseudo-Ciceronian style.[43] It had already fallen out of favor by the beginning of the seventeenth century, and so this type of style is for Wilkins a side note. The second kind of rhetorical affectation proves more complicated, more dangerous to the good order of society, and far more pervasive—"much in fashion." The "mystical kind of phrase" to which Wilkins refers is the language of the sorcerer and the enthusiastic spiritualist. Such rhetoric "signifie[s] nothing" for the Anglican bishop, except perhaps a pact with the Devil, as in the case of Macbeth's troubled rhetoric. Wilkins conjunction of "signifie nothing" and "vain repetitions" (e.g., double, double) might very well be an obscure reference to Shakespeare's meditation upon how demons and witches infect the human idiom. Even if the allusion seems strained, Wilkin's broader message is unmistakable: many Interregnum writers suffer a nefarious influence in their rhetoric, as evidenced by their applications of mystical phrases not to be found "in Scripture." Such phrases sound sacred, but, Wilkins suggests, are "wholie empty," subverting divine eloquence under the guise of promoting it, in what amounts to an act of rhetorical topsy-turvydom.

Style is a dominant philosophical theme throughout the *Discourse*, not only as it relates to prayer and sermonizing, but also as it relates to communication in general. The book's title is probably misleading, as Francis Christianson points out, because Wilkins argues for plainness in all areas of writing. He takes an interest in discourse broadly imagined. In fact, Christianson goes even farther: "There is little, if anything, in the resolutions of the Society concerning their discourse that is not already formulated here by Wilkins."[44] Such a claim

42 John Wilkins, *A Discourse Concerning the Gift of Prayer* (London, 1653), 48.

43 Croll, *Style, Rhetoric, and Rhythm*; Robinson, *The Establishment of Modern English Prose*, 142–143.

44 Christianson, "John Wilkins and the Royal Society's Reform of Prose Style," in *John Wilkins and 17th-Century British Linguistics*, ed. Joseph L. Subbiondo (Amsterdam: Benjamins, 1992), 151.

is probably too bold, because Sprat's plain style mandate is more developed than Wilkins's, especially as his philosophy of rhetoric appears in this early book. Christianson is nonetheless right to notice Wilkins and Sprat's shared interest in language reform. After all, Wilkins nominates Sprat to write the *History*, due in large part to their close friendship, not to mention their commensurable attitudes toward rhetoric.

Wilkins advances a similar argument for plainness and against rhetorical sorcery in a sermon upon Ecclesiastes:

> Words being but the images of matter, and to be wholly given up to the study of these, what is it but *Pygmalions* phrenzy, to fall in love with a picture or image, as for *Oratory* which is the best skill about words, that hath by some Wisemen been esteemed but a voluptuary art, like to cookery, which spoils wholesome meats, and helps unwholesome, by the variety of sauces serving more to the pleasure of taste than the health of the body.[45]

The Bishop draws a contrast between the experimentalist's attitude toward words and the magician's, paraphrasing *The Advancement of Learning* in the process—Bacon's Pygmalion argument.[46] And like Bacon, Wilkins critiques the occult obsession with rhetorical charms. Extending the argument, Wilkins also provides the commonplace metaphor associating bad rhetoric with bad cookery. In the same way that devious cooks disguise unhealthy food with savory sauces, so too do misguided orators make the worse appear the better reason, he suggests, disguising the bitter as the sweet. Analogically, the pot roast is the substance in this argument, while the sauce functions as ornament, which might give temporary pleasure, but affords no sustenance and in fact proves dangerous. Wilkins's remark about "odd humors" is especially interesting along these lines. The implication of this quasi-medical phrase is that certain unwholesome forms of rhetoric (sauces) corrupt the imagination and the nervous system, transporting people to dangerous places by disturbing the humors. For Wilkins, the problem of venomous rhetorical cookery is a serious one, as serious as venomous cookery itself. Indeed, the persistence of poisonous rhetoric explains for many Anglican philosophers how the English Civil War began. That is, intoxicating demonic eloquence—nonconformist eloquence—moved a lot of people falsely. Vapors (idioms) of the underworld

45 Wilkins, *Sermons Preached upon Several Occasions* (London, 1682), 184–185.

46 In *The Advancement of Learning*, Bacon uses the image of Pygmalion as an emblem of "the first distemper of learning, when men study words and not matter," with the caveat that magicians, astrologers, and alchemists are especially guilty of this offense (52).

overwhelmed a large congregation of Englishmen (i.e., disturbed the humors), which caused them to abandon Anglican orthodoxy and new science, preferring instead the path of mystical eloquence, which by latitudinarian standards poisoned the rational faculties and moved the will badly.

Using the Pygmalion story in a sermonic setting, Wilkins also warns of rhetorical apostasy, especially in the behaviors of Christians who suffer from overly reverential attitudes toward relics, which Aquinas categorizes much earlier as a dangerous form of idolatry.[47] Wilkins concerns himself with the worshipping of false idols, including the *materia* and *forma* of tropes. At the same time, he is unconcerned about the use of eloquence, as is evidenced by his own spectacular rhetoric. The turn of phrase that troubles Wilkins is a preternatural turn, an act that destabilizes learning's advancement and the concomitant Anglican theology of style contained therein.

Typical of Anglican rhetorical theory at the time, Wilkins also devotes attention to the related topics of iterative preaching and enthusiastically spontaneous preaching, warning against both practices in the *Discourse Concerning the Gift of Preaching* (1646):

> [Preaching] must be full, without empty and needless Tautologies, which are to be avoided in every solid business, much more in sacred. [And] to deliver things in a crude and confused matter, without digesting of them by previous meditation, will nauseate the hearers, and is as improper for the edification of the minde, as raw meat is for the nourishment of the body. It must be sound and wholesome, not tainted with any erroneous corrupt doctrine, or the affectation of novelty. False opinions do many times insinuate themselves by the use of suspicious phrases. And 'tis a dangerous fault, when men cannot content themselves with the wholesome sound of words, but do altogether affect new light and new language, which may in time destroy practicall Godlinesse and the power of Religion.[48]

First, the Bishop rejects "empty and needless Tautologies," which his audience would have immediately apprehended as anti-Catholic, and they would have expected as much.[49] Wilkins's purpose is to differentiate between orthodox

47 Aquinas, *Summa theologica*, III: 25:6.

48 Wilkins, *A Discourse Concerning the Gift of Preaching* (London, 1646), 129.

49 On Catholicism in seventeenth-century England, see Raymond Tumbleson, *Catholicism in the English Protestant Imagination. Nationalism, Religion, and Literature 1600–1745* (Cambridge: Cambridge University Press, 1998).

Anglican ceremony and alternative forms of iteration. While the former is marked by a plain attitude toward rhetoric and an "experimentall acquaintance" with truth, the latter appears unabashedly mystical.[50] These troubled incantations, moreover, bring with them the possibility of demonic ritual disguised as true religion, a commonplace concern among theologians on all sides of the religious controversies, and a charge that the Anglicans regularly leveled against Catholic and Lutheran ceremonies.

On the related topic of novel and spontaneous rhetoric, Wilkins has in mind Puritan preaching. The Puritans disavowed formulaic religious rhetoric, arguing that such practices were too much aligned with the apparatus of witchcraft. This included both Anglican and Catholic ceremonies. For the Puritan sensibility, set formulas interfered with the direct inspiration of the Holy Spirit, and herein lies one of the major reasons why the Puritans preferred Ramus to classical rhetoric.[51] They perceived in the classical canon of invention an artificiality that worked against God's moving grace. Overly prescriptive heuristic procedures circumvented inspiration. Additionally, and of equal significance to the Puritans, classical philosophers took the audience's disposition too seriously. It was the audience's responsibility to the Puritan preacher to be prepared to receive God's wisdom, not the preacher's responsibility to adjust the message to the audience. All of this is by way of suggesting that Bishop Wilkins confronts the nonconformist challenge to Anglican orthodox rhetoric, which manifests itself through unabashedly mystical theology and spontaneous eloquence, or what Milton referred to as "prompt eloquence" in *Paradise Lost.*

On a side note (if I might digress for a moment), we discover in the Puritan rejection of classical heuristics the seeds of English Romanticism. The Romantic's renewed interest in Cromwell and Interregnum rhetoric makes perfect sense from a philosophical-rhetorical standpoint, because the nonconformists foreshadow Romantic ideas of inspiration, not to mention Romantic ideas of science, where theosophy plays a crucial role. Enlightenment philosophers such as Bishop Burnet and John Toland represented as superstitious certain key aspects of Puritan *furor*, but Romantics like Wordsworth, Shelley, and Keats took great encouragement from such charmed rhetoric. They embraced a mystical element in composition, celebrating enchanted modes of invention and inspiration, and therefore finding more in common with Renaissance mysticism than with Enlightenment science.

---

50 Wilkins, *A Discourse Concerning the Gift of Preaching*, 130.

51 On Ramistic rhetoric, see Ong, *Ramus.*

Archbishop John Tillotson is perhaps the most influential writer on the topic of sermonic style in Restoration England.[52] A member of the Royal Society and a Baconian philosopher, he defends the plain language reforms with the same vehemence as Wilkins and Sprat. The new "plain" style of preaching, and of writing more broadly, Tillotson argues,

> may correct the irregular humor and itch in many people who are not contented with this plain and wholesome food, but must be gratified with sublime notions and unintelligible mysteries, with pleasant passages of wit, and artificial strains of rhetoric, with nice and unprofitable disputes, with bold interpretations of dark prophecies, and peremptory determinations of what will happen next year, and a punctual stating of the time when antichrist shall be thrown down.[53]

Tillotson contrasts properly plain language with mystical and ecstatic rhetoric. Among the types of *esoterica* at issue, he identifies those "bold interpretations" of "dark prophecies" that determine punctually "when antichrist shall be thrown down." The Millenarians are the most notorious of the nonconformist apocalyptic philosophers of the Interregnum, and Tillotson no doubt has this group's mystifying rhetoric in mind, at least in part.[54] Millenarian apocalyptic expressions were commonplace in seventeenth-century eschatological writings (e.g., Joseph Mede, James Durham, John Napier), despite the fact that this sort of formulaic thought worked against the spirit of that most famous admonition in Matthew 25:13: "Watch therefore, for ye know neither the day nor the hour the Son of man cometh." For some of the more fervent eschatologists, this warning remained true only in a technical sense, with the emphasis upon day and hour. Tillotson undoubtedly refers to these adamant Millenarian discourses when he complains about "unintelligible mysteries" and "artificial strains of rhetoric."

52 On Tillotson's influence upon Restoration style, see Adolph, *The Rise of Modern Prose Style*. See also Mary Morrissey, "Scripture, Style, and Persuasion in Seventeenth-Century English Theories of Preaching," *Journal of Ecclesiastical History* 53 (2002), 686–706; Gerard Reedy, "Interpreting Tillotson," *Harvard Theological Review* 86 (1993), 81–103; David Brown, "John Tillotson's Revisions and Dryden's 'Talent for English Prose,'" *Review of English Studies* 1961 (45), 24–39.

53 Tillotson, "The Necessity of Repentance and Faith," in *The Works of the Most Reverend John Tillotson*, 3 vols. (London, 1717), vol. 2, 11–17, 6.

54 On the Millenarians, see Bryan Ball, *A Great Expectation. Eschatological Thought in English Protestantism to 1660* (Leiden: Brill, 1975), 55–88; Norman Cohn, *The Pursuit of the Millennium* (Fairlawn: Essential Books, 1957).

The Millenarians, however, represent a small element of the seventeenth-century writers who took seriously the idea of an impending apocalypse. Not only does Tillotson target the peculiar pronouncements of Millenarians, who play a relatively minor role in debates about providential history, but he also targets the far more widespread eschatological sentiments of mainstream Anglican, Lutheran, Catholic, and Puritan theologians. A majority of theologians, conventional and radical, conformist and nonconformist, considered the real possibility of an imminent apocalypse in the middle of the seventeenth century. It is difficult to overstate the extent to which eschatology influenced English theology and rhetoric, beginning in earnest with the Reformation and extending into the Interregnum. As Bryan Ball shows, the belief in an imminent apocalypse was "nearly universal" among English Reformists: "At no other time in England's history has the doctrine of the second advent been so widely or so readily accepted."[55] Stuart Clark makes a similar argument: "The advent of the Antichrist [in the form of the Papacy] was taken as the surest of many signs that the English Reformation was a part of that decisive battle between good and evil as described in Revelation."[56] Writers from various Protestant standpoints believed that the age of Antichrist had arrived on earth in the form of Catholic empire, an idea illustrated in Luther's theology and Melanchthon and Cranach's hugely popular *Passional Christi und Antichristi* (1521). This same apocalyptic anxiety emerges in foundational English Protestant texts such as John Bale's *The Image of Bothe Churches* (1545), Thomas Cranmer's *Catechismus* (1548), John Cheke's *The Hurt of Sedition* (1549), Thomas Wilson's *Arte of Rhetorique* (1553), and Hugh Latimer's *Sermons* (1562). For many Catholics, too, the idea of the day of reckoning loomed large. The Reformation itself operated in the Catholic Counter Reformation as evidence of *postremus furor Satanae*, Satan's final rage, where the Devil attempts to cause as much heresy as possible before God banishes him forever to hell—an idea based upon Revelation 12:12: "Woe to the inhabiters of the earth and of the sea, for the devil is come down unto you, having great wrath, because he knoweth he hath but a short time."

When Tillotson takes up the issue of apocalyptic rhetoric, he invokes this complex theological milieu that permeates the period's religious debates. In one strong gesture, too, he dismisses most of the apocalyptic tradition as belonging to a credulous world of arcane signs and symbols that preceded

55 Ball, *Eschatological Thought in English Protestantism to 1660*, 232.

56 Stuart Clark, *Thinking with Demons* (Oxford: Clarendon Press, 1997), 339. On rhetoric and apocalypse in Renaissance England, see Ryan J. Stark, "Thomas Wilson's Apocalyptic Rhetoric," *Studies in Philology* 106 (2009), 341–353; Stark, "Protestant Theology and Apocalyptic Rhetoric in Roger Ascham's *The Schoolmaster*," *Journal of the History of Ideas* 69 (2008), 517–532.

the world of learning's advancement. The cosmos had changed by the time Tillotson inherited the apocalyptic tradition. Or, more precisely, he and other new philosophers were in the process of changing the cosmos. Radical eschatological sentiments no longer seemed tenable, and the idiom in which they were written appeared increasingly obscure as the plain tropes of new philosophy took hold in English historiography, moving historiography closer to a progressive model and away from an Augustinian binary model, marked by the metaphysical and absolute struggle between good and evil in every epoch.[57]

This is not to say that Tillotson rejects the basic idea of eschatology. He does not, and neither do other members of the Royal Society. The book of Revelation is an important text in the late seventeenth century, as are other apocalyptic writings, but they are not as central. Like most Restoration philosophers working in the tradition of modern experimentalism, Tillotson perceives in the present age the dawn of a new golden era, where advances in science and art signal a major step forward in human development. Scientific optimism counteracts the twilight world of apocalypse that many English writers articulated consistently throughout the Renaissance. As a result, apocalyptic rhetoric suffers a profound loss in momentum in the late seventeenth century. The plain language reforms contribute significantly to this loss, because the new philosophers' plain idiom contains as a presupposition a profound misgiving toward eschatological portents and symbols, including astrological portents, which is to say that the decline of academic astrology and the rise of the new plain style are intimately related. While apocalyptic rhetoric finds expression in every age, it becomes in the Enlightenment an increasingly peripheral practice. Perhaps a work like Pope's *Dunciad* provides a counterpoint, especially the fourth book. But in Pope's mock epic, the apocalypse functions as a literary device more than a real possibility. Colley Cibber is the Antichrist of wit, but not by any stretch is he the Antichrist of the universe. Pope warns England about reason's collapse, but his satirical method dissipates the sober idea of the world's epilogue, which as a concept seems far removed from the Age of Reason and Pope's balanced couplets contained therein.

The apocalyptic rhetoric of Renaissance theology appears increasingly as a period style to advancers of learning, a set of tropes and ideas best suited to a particularly enthusiastic stretch of time that the Restoration Anglicans desired to obscure through an official Act of Oblivion and other efforts to start anew with socio-historical *tabulae rasae*. The rhetoric of apocalypse begins to fade in

57 On this shift in historiography, see Blair Worden, "The Question of Secularization," in *A Nation Transformed. England after the Restoration*, eds. Alan Houston and Steve Pincus (Cambridge: Cambridge University Press, 2001), 20–40.

the Restoration, at least in mainstream Anglican circles, and this includes Tillotson's circle.

Tillotson also packs into the passage above another intriguing trope, and this is the commonplace metaphorical distinction between the "wholesome food" of good rhetoric and the unwholesome food of bad rhetoric. Although a full survey of food metaphors and rhetoric in the seventeenth century has yet to be completed, the metaphor of bad rhetoric as bad food is probably the third most commonly used trope to disparage inappropriate styles in the middle of the seventeenth century, behind bad rhetoric as witchery and bad rhetoric as harlotry, two manifestations of demonic inversion. Significantly, all three of these images: witches, harlots, and bad cooks, have in common the bedrock association of the demonic, though the bad food metaphor is less overt than the others. Nonetheless, embedded deeply in the routine references to unwholesome food and bad rhetoric are the dual and related Anglican anxieties over sophistry's pharmacy and Catholicism's transubstantiation. The first, the pharmacy, goes back at least as far as Plato's critique of Gorgias. Plato, Aristotle, Augustine, and other inventors of the rhetorical tradition held that certain kinds of styles functioned as dangerous drugs, corrupting the psyches and nervous systems of audiences and orators alike, especially if consumed uncritically. This is a sensible belief, then and now. Some forms of rhetoric poison the system. In this light, Tillotson's phrase "irregular humor and itch" sounds like a medical diagnosis, and it probably is. He suggests that consumers of enchanted rhetoric put themselves in physical danger by virtue of their perverse appetites, or, worse yet, have already been corrupted by the drug of diabolical language. In either case, the outcome is deleterious. Bad rhetoric, like bad food, perverts healthy living.

The second connotation of unwholesome food that lingers in the background of Tillotson's metaphor is that of Catholic transubstantiation. In Catholic doctrine, the bread and the wine of communion literally turn into the body and the blood of Christ. The Catholic Council of Lateran in 1215 articulated the details of this position, just as the mystical rhetoric of the early Italian Renaissance burgeoned. The accidental properties of taste and texture persisted in the bread and wine, as the doctrine indicates, but in a deeper sense, though still in a literal sense, the very nature of the sacraments change into the actual body and blood of Christ. Moreover, transubstantiation occurs at that very moment when the priest utters sacred words. This makes communion the most prominent of all mystical (not magical) rhetorical ceremonies in Catholicism, and consequently the most prominent of all mystical rhetorical ceremonies in European Renaissance society. To Tillotson's Anglican sensibility, however, the Catholic ritual of communion appeared as one more form of nefarious preternatural magic, indistinguishable in category from the rhetorical behaviors of necromancers, alchemists, and witches. Unlike the Catholics, the Anglicans took a purely symbolic view of communion, while the Lutherans found a middle ground, perceiving in the sacrament consubstantiation, not transubstantiation. All of this is by way of preface to say that Tillotson alludes to Catholic communion when he describes those who "must be gratified with sublime notions and unintelligible mysteries," instead of nourishing food. The mystery of transubstantiation is for Tillotson and other latitudinarian philosophers one more example of an "unintelligible [rhetorical] mystery," and this is due to the mystical philosophy of rhetoric at work in the ceremony, which strikes the Anglican sensibility as a form of fiendish metonyming.[58] Against enchanted rhetorical remedies of all sorts, Tillotson juxtaposes the new plain style, a wholesome form of rhetoric by the standards of his theologically minded food analogy.

Not all bad-rhetoric-as-bad-food metaphors in the middle of seventeenth-century England involve demonry, occult pharmacology, or transubstantiation, but many do, at least on the subtextual level and from the standpoint of the Royal Society's mainstream philosophers. The Anglican Henry More provides another illustration, invoking a food metaphor with occult overtones: "Enthusiasm," he argues, is "an affection of Humor and Rhetorick," and it is marked by a "desire to be filled with high-swoln words of vanity, rather than to feed on sober truth, and to heat and warm our selves rather by preposterous and fortuitous imaginations, than to move cautiously in the light of a purified mind and improved reason."[59] Consuming the food of bad rhetoric puts one at risk of having one's perceptions turned upside down. More's use of the term "preposterous" is especially notable here, because it recalls the rhetorical inversions of witchcraft, which he aims to connect with nonconformist ranting. Put differently, More complains about those English enthusiasts who feed on honey dew and drink the milk of paradise. His advice about producing and consuming such imaginative rhetoric is simple: Beware! Beware! Cowley provides a similar warning, comparing bad rhetoric to bad food in his ode to the Royal Society. Both substances produce "Pageants of the brain," rather than the sound rationality that comes from proper rhetorical sustenance (e.g., the new plain style).[60] Notably, too, Cowley invokes an upside-down universe—the pageant—as a part of his broader rejection of diabolical rhetoric, in this case juxtaposing the perverse consumption of the Renaissance masquerade with

58 See also Tillotson, *A Discourse against Transubstantiation* (London, 1684).

59 Henry More, *Observations upon Anthroposophia Theomagica* (London, 1650), A3v.

60 Cowley, "To the Royal Society," in *The History of the Royal Society*, Preface.

the proper consumption of new science. Using the striking antithesis, he also reveals the Royal Society's efforts to calm social misrule through the healthy influence of plain rhetoric. This is Tillotson's point as well: the alternative to debauched rhetorical nourishment, or demonic rhetoric, is the purified food of new philosophy, which finds its analogue in the Royal Society's plain language program. Plainness heals, Tillotson contends, while esoteric rhetoric corrupts physiological, social, and spiritual systems.

In his *Discourse of the Pastoral Care* (1692), Bishop Gilbert Burnet also puts magic and mysticism at the crux of the plain language controversies, noting how mystical rhetoric threatened "plain Notions" of preaching in the Interregnum and the early Restoration:

> Mystical Applications of Scripture grew to be better liked than clear Texts; an Accumulation of Figures, a Cadence in the Periods, a playing upon the Sound of Words, a Loftiness of Epithets, and often an Obscurity of Expression, were according to the different tastes of several Ages run into. Preaching has passed through many different forms among us since the Reformation. But without flattering the present Age, or any person now alive, too much, it must be considered, that it is brought of late to a much greater Perfection than it ever was before among us. It is certainly brought nearer the Pattern that S. Chrysostom has set, or perhaps carried beyond it. Our Language is much refined, and we have returned to the plain Notions of simple and genuine Rhetorick.[61]

Burnet takes bad mysticism to be the root cause of England's rhetorical and social problems. The symptomatic and secondary troubles of cadence and obscurity flow from it. And the cure, as he figures it, is to adopt plain notions of rhetoric, not esoteric notions.

Burnet preached Robert Boyle's and Tillotson's funeral sermons, and he was a longstanding member of the Royal Society. It is fair to say that he participated firsthand in the plain language reforms. Philosophers of rhetoric, therefore, do not get an impartial report from him, but we discover a coherent rationale for the plain style movement from the standpoint of the plain language reformers themselves. By the time Burnet wrote the *Discourse*, this rationale against rhetorical enchantment rang true in the society at large. On the cusp of the eighteenth century, the darksome rhetoric of the old world was giving way to the polite, cool, plain rhetoric of the new epoch, at least in typical intellectual circles, which then spread the new philosophy of plainness to the increasingly

61 Gilbert Burnet, *A Discourse of the Pastoral Care* (London, 1692), 215–216.

large and literate public. Ideas of mystical metonyms and occult paronomasias faded in latitudinarian discourses. Astrology, magic, witchcraft, and other esoteric systems of thought, of course, persisted in eighteenth-century English culture, and beyond, and these systems still have numerous high-profile supporters today, as enchanted philosophy does in every age. It would be a mistake, therefore, to talk about the collapse of charmed rhetoric in an unqualified way. In the context of Burnet's treatise, however, illuminists and magicians no longer posed the strong challenge to the new experimentalism that they did in the early and middle parts of the seventeenth century. Mystical and magical worldviews were no longer central topics in academic discussions about truth and reality, as they were, for instance, during the Royal Society's inception of plainness. Those advocating rhetorical plainness (i.e., non-enchanted stylistics) won the philosophical battle, more or less, not in an absolutist Hegelian sense, where the innovative *Zeitgeist* replaces the old *Zeitgeist* (and the new period style eliminates the contingencies of the old period style), but in a Thomas Kuhnian sense of paradigm shift, where the old paradigms remain thoroughly intact but lose considerable esteem in mainstream intellectual society, for better or worse.[62]

Burnet's optimism, too, is notable. The new philosophy of plainness has "carried beyond" even the best rhetoric of antiquity. For Burnet, the modern plain style is poised to usher in a new era—an Age of Reason. The Bishop anticipates this turn. Indeed, he advances it, celebrating plainness and perspicuity in preaching, both of which work against the metaphysical conceits and occult verisimilitudes of early seventeenth-century religious discourses. This feeling of progress in Burnet also and significantly contradicts the end-of-the-world scenarios posited by the apocalyptic philosophers, those who believed that Antichrist had arrived on earth, which meant that the Second Coming would not be far behind. In a striking counterstatement to end-of-the-world angst (including those "peremptory determinations" discussed by Tillotson), Burnet sees the front cusp of a golden age where language is refined to "a much greater Perfection than it ever was before." He detects the progress of learning,

62 Thomas Kuhn, *The Structure of Scientific Revolutions* (Chicago: University of Chicago Press, 1962). See also Margaret Osler, "Rethinking the Scientific Revolution," in *Rethinking the Scientific Revolution*, ed. Osler (Cambridge: Cambridge University Press, 2000), 3–24; Amos Funkenstein, *Theology and the Scientific Imagination from the Middle Ages to the Seventeenth Century* (Princeton: Princeton University Press, 1986), 346–360; Ian Hacking, *Representing and Intervening. Introductory Topics in the Philosophy of Natural Science* (Cambridge: Cambridge University Press, 1983), 65–75; Hacking, *The Emergence of Probability* (Cambridge: Cambridge University Press, 1975).

not civilization's collapse. That is, Burnet looks forward, as do the other plain language reformers of the late seventeenth century.

Conclusion

Near the end of *The Establishment of Modern English Prose*, Robinson comments upon Oliver Cromwell's mystical rhetoric. Referring to a heartfelt letter that Cromwell writes about the death of one of his soldiers, Robinson makes the following point about the Protector's writing style:

> The Lord's name is frequently on Cromwell's lips, for which A.L. Rowse (for instance) can hardly mention him without accusing him of hypocrisy. [Cromwell's writing] doesn't sound like the style of a hypocrite. But nor does it sound quite like modern prose, though not for any technical reason. Neither unembarrassed emotion nor a sense of the presence of God are associated with modern prose. Cromwell is a world away from Restoration prose, a thousand miles from the coffee house and quite without the bows and scrapes Addison manages to set down on paper. The connection with the Bible is the reverse of a weakness. Cromwell really did pant for happiness in the Lord! No writer at the court of Charles II could have used glory so without embarrassment and so convincingly.[63]

This passage is the theoretical monad of Robinson's book, even though he presents it as an afterthought, probably because his interpretation in the passage departs severely from the rhythmic-syntactical approach to seventeenth-century stylistic reform governing most of the book. As Robinson notes, Cromwell does not sound like a modern writer, "though not for any technical reason." In other words, Cromwell's style is not really plain or perspicacious in the modern sense, even though it appears that way technically. What might this mean? Cromwell's rhetorical style still participates in a mystical world where the Spirit of God informs daily living, including daily writing, and this mystical aura comes through in the language, especially when he uses terms like "glory" with neither embarrassment nor Restoration restraint. We discover in Cromwell an excellent example of nonconformist spirited rhetoric, which has more in common with metaphysical and occult temperaments than with the plain temperaments of the new experimentalists. Donne and Cromwell,

63 Robinson, *The Establishment of Modern English Prose*, 138.

for example, are much closer to each other than either writer is to Wilkins, Sprat, or Locke. Cromwell's eloquence, in fact, functions as one of several foils for the rise of the new plain style. Members of the Royal Society and other like-minded philosophers successfully targeted such numinous sensibilities, building a modern style against them.

Robinson foreshadows his discussion of Cromwell by noting that John Bramhall's rhetoric might very well be construed as the forerunner of Dryden's proto-Enlightenment style, given Bramhall's masterful ability to write prose in a way that approximates ordinary speech. At the same time, as Robinson argues, "Bramhall has a genuine religious feeling that divides him decisively from Dryden."[64] Robinson is right. Moreover, it is precisely this sort of thing—a particular type of religious feeling—that philosophers of rhetoric need to articulate more clearly in such circumstances, if the language reforms of the early modern period are to be properly conceptualized, and if the emergence of modern English writing is to be apprehended for what it is: a movement against enchanted rhetoric in all of its forms, mystical and magical. For the intuitive critic who apprehends not only the shape of the sentence but also the shape of the world in which the sentence functions, the aura of Bramhall's writings is easily distinguishable from the aura of Dryden's writings, not necessarily because of syntactical differences, but because of thermal-mystical differences. On the level of mere sentence structure, Bramhall has far more in common with Dryden than he does with Browne or Vaughan, but such a syntactical focus proves entirely misleading. Put simply, there is a numinous warmth in Bramhall that does not exist in Dryden's austere writings.[65] How might philosophers of rhetoric discover such warmth in a case like this? Perhaps a hermeneutical procedure implied in Robinson's discussion of Cromwell is a good place to start. Can the author say the word "glory" without embarrassment?

We need an understanding of a sentence's *Weltanschauung* in order to determine its stylistic implications, in addition to a more easily discoverable syntax gauge that measures types of clauses. A syntactical gauge in itself is of very little use, and has produced much confusion in rhetorical studies. In the absence of a broader philosophy of rhetoric, the shape of syntax, use of particular tropes, and proclivity for certain types of clauses tell us very little about a writer's attitude toward style and, concomitantly, toward the world.

64 *Ibid.*, 122.

65 I am thinking of Dryden's writings before 1685, though the Catholic Dryden remains a rationalist and an experimentalist in the tradition of Descartes.

CHAPTER 4

Argument and 17th-Century Science

*A Rhetorical Analysis with Sociological Implications**

Alan G. Gross, Joseph E. Harmon and Michael S. Reidy

It is now over two decades since Bruno Latour and Steve Woolgar set up their tent amid the inhabitants of the Salk Institute, La Jolla, California: "a strange tribe who spend the greatest part of their day coding, marking, altering, correcting, reading, and writing."[1] Part anthropology, part sociology and part rhetorical analysis, their *Laboratory Life* vividly portrayed how social interactions enter into the making of "literary inscriptions" the authors hope will enter the domain of accepted knowledge among their peers.[2] Since then, many other studies have demonstrated that rhetorical analysis can lead to conclusions relevant to the sociology of science.[3] We believe that as rhetoric becomes even

* This research was supported at various times by grants from the University of Minnesota and the Society for Technical Communication, and was accomplished in part while the first author was in London as a British Association Visiting Professor as well as a Fellow at the Center for the Philosophy of Science, University of Pittsburgh; the Institute for Advanced Studies, Hebrew University; and the Internationales Forschungszentrum Kulturwissenschaften. The contribution of the second author was performed independently of his institutional affiliation. We thank the journal referees for their astute and constructive comments on the original manuscript. Photographs courtesy of University of Minnesota libraries.

1 Bruno Latour and Steve Woolgar, *Laboratory Life. The Construction of Scientific Facts* (Princeton, NJ: Princeton University Press, 2nd edn, 1986), 49.

2 Two other early sociological studies involving analysis of scientific texts are G. Nigel Gilbert and Michael Mulkay, *Opening Pandora's Box. A Sociological Analysis of Scientists' Discourse* (Cambridge: Cambridge University Press, 1984), and Karin D. Knorr-Cetina, *The Manufacture of Knowledge. An Essay on the Constructivist and Contextual Nature of Science* (Oxford: Pergamon Press, 1981). Among other insights, these studies found that the modern scientific article represents a rational reconstruction of scientific work, one that distorts the true nature of that work and suppresses some of its underlying motivations.

3 Our review of the last 15 years' worth of *Social Studies of Science*, for example, turned up the following studies with a strong rhetorical slant. Charles Bazerman, "Modern Evolution of the Experimental Report in Physics. Spectroscopic Articles in *Physical Review*, 1893–1980," *Social Studies of Science* 14, 2 (May 1984), 163–196; Greg Myers, "Texts as Knowledge Claims. The Social Construction of Two Biology Articles," *ibid.* 15, 4 (November 1985), 593–630; Myers, "Politeness and Certainty. The Language of Collaboration in an AI Project," *ibid.* 21, 1 (February 1991), 37–73; Myers, "From Discovery to Invention. The Writing and Rewriting of Two Patents," *ibid.* 25, 11 (February 1995), 57–105; Bryce Allen, Jian Qin

 | DOI 10.1163/9789004283701_006

further integrated as a method within the sociology of scientific knowledge, more uses will be found for it and, as a result, scientific communities will more often be seen as communities whose communicative and persuasive practices constitute necessary conditions to the creation of new knowledge.

To add some weight to this general case, we have examined the argumentative practices of the three key periodicals reporting 17th-century science: the *Philosophical Transactions, Journal des Sçavans* and *Mémoires de l'Académie Royale des Sciences.* These three have an important common social bond: each has some connection with a scientific society—the Royal Society of London for *Philosophical Transactions* and the Académie Royale des Sciences of Paris for both the *Journal des Sçavans* and *Mémoires.* These neophyte organizations had within their ranks most of the authors, readers and journal editors of 17th-century science. They also created the social networks needed to establish what constituted acceptable argumentative practices in science.[4]

and F.W. Lancaster, "Persuasive Communities. A Longitudinal Analysis of References in the *Philosophical Transactions* of the Royal Society, 1665–1990," *ibid.,* 24, 22 (May 1994), 279–310; Benoît Godin, "The Rhetoric of a Health Technology. The Microprocessor Patient Card," *ibid.,* 27, 6 (December 1997), 865–902; Harro van Lente and Arie Rip, "The Rise of Membrane Technology. From Rhetorics to Social Reality," *ibid.,* 28, 2 (April 1998), 221–254. In the first of these papers, Bazerman investigated the changing character of communication in 20th-century physics. In three separate papers, Myers analyzed the negotiation of the breadth of biological claims during peer review, the function of politeness strategies in scientific decision-making, and the status of knowledge claims in patents as distinct from scientific papers. Allen, Qin and Lancaster examined citation patterns in the *Philosophical Transactions* as a window on to the persuasive practices of scientific communities over time. Moving in a different direction, van Lente and Rip integrated rhetorical analysis into their examination of the creation of a new social reality in "membrane technology," and Godin proceeded similarly in tracing the social history of the Canadian microprocessor patient card. Although the evidence from the dates of these articles is no more than suggestive, and hardly constitutes a trend, it does suggest a growing interest in the rhetoric of science and technology: 1984, 1985, 1991, 1994, 1995, 1997, 1998.

4 There are many excellent historical works on the early journals and scientific societies, including Sherman B. Barnes, "The Scientific Journal, 1665–1730," *The Scientific Monthly* 38 (1934), 257–260; Raymond Birn, "Le journal des savants sous l'Ancien Régime," *Journal des Savants* (Janvier-Mars 1964), 15–35; Roger Hahn, *The Anatomy of a Scientific Institution. The Paris Academy of Sciences, 1666–1803* (Berkeley: University of California Press, 1971); Marie Boas Hall, "The Royal Society's Role in the Diffusion of Information in the Seventeenth Century," *Notes and Records of the Royal Society* 29 (1975), 173–192; David A. Kronick, *A History of Scientific & Technical Periodicals. The Origin and Development of the Scientific and Technical Press 1665–1790* (Metuchen, NJ: Scarecrow Press, 1976); Hub Laeven, *The* Acta Eruditorum *under the Editorship of Otto Mencke (1644–1707). The History of an International*

No two groups founded at nearly the same time with similar interests could be more different in organization and constituency than the Royal Society and the Académie Royale.[5] The Royal Society was a fairly large, loose-knit group of amateurs and "natural philosophers" in and around London—some extraordinarily talented, some with nothing more extraordinary than an above-average curiosity about the natural world. They met regularly to learn about and discuss the latest experiments and observations, as well as to witness demonstrations of them. The Society received the blessing of the reigning monarch, Charles II, but depended largely upon members' dues and generosity for its economic survival.[6] The Royal Society secretary, Henry Oldenburg, launched *Philosophical Transactions* in March 1665 as a private venture. His transactions reported on technical news from home and abroad, often sent to him in the form of letters from his many correspondents.[7]

In contrast to the Royal Society, the Académie Royale was closer to what we think of today as a government-funded research institute. Science was the principal occupation of its members, who lived and worked together in Paris under the patronage of Louis XIV, funded "at a level similar to the annual income of the wealthiest monastery in France."[8] The Académie was also deeply involved in the publication of the fruits of its members' labours under the Imprimerie Royale. During the 17th century, it published two handsome volumes on plants and animals, as well as a large collection of articles on mathematics and physical science, by "messieurs de l'Académie Royale." The first scientific 'mémoires' of the Académie Royale date from 1666, but most of these

---

Learned Journal between 1662–1707 (Amsterdam: Holland University Press, 1990); Roger Philip McCutcheon, "The *Journal des Sçavans* and the *Philosophical Transactions* of the Royal Society," *Studies in Philology* 21 (1924), 626–628; James E. McClellan, III, *Science Reorganized. Scientific Societies in the Eighteenth Century* (New York: Columbia University Press, 1985); Douglas McKie, "The Scientific Periodical from 1665 to 1798," *Philosophical Magazine*, Commemorative Issue (1948), 122–132; Martha Ornstein, *The Role of Scientific Societies in the Seventeenth Century* (Chicago, IL: The University of Chicago Press, 2nd edn, 1928); Dorothy Stimson, *Scientists and Amateurs. A History of the Royal Society* (New York: Greenwood Press, 1968); Alice Stroup, *A Company of Scientists. Botany, Patronage, and Community at the Seventeenth-Century Parisian Royal Academy of Sciences* (Berkeley: University of California Press, 1990).

5 From work cited in note 4, see Hahn, 1–34; Hall; McClellan, 41–66; Stroup, 3–61.

6 McClellan (*op. cit.* note 4, 30) notes that in the 18th century the Royal Society did receive governmental funding for "several [large] expeditions and projects that had an immediate bearing on the government's interests in navigation, trade, and colonial expansion."

7 Henry Oldenburg, *Correspondence*, ed. and trans. A. Rupert Hall and Marie Boas Hall (Madison: University of Wisconsin Press, 1965–1986).

8 Stroup, *op. cit.* note 4, 34.

were not actually published until the early 18th century in the multivolume *Mémoires de l'Académie Royale des Sciences, depuis 1666 jusqu'à 1699*. However, articles by Académie members and summaries of their research did regularly appear throughout the 17th century in *Journal des Sçavans,* started privately in January 1665 by Denis de Sallo.

Joseph Ben-David captured a key social difference between these two early scientific societies:

> [W]hereas the Royal Society was an independent corporation based on membership including amateurs and politicians of science as well as scientists of outstanding accomplishments, the *Académie* was a sort of elevated civil service composed of only a small number of scientists of high reputation.

Importantly, the French arrangement had as its purpose

> ...to control science and limit its influence.... [T]he empirical and experimental approach of science [was not to be] diffused to politics...the norms of universalism of science [were not to be] applied to matters of religion and social estate.[9]

These social differences also had, as we hope to demonstrate, consequences that affected argumentative practices as manifested in the concrete utterances found in journal articles.

Methodological Considerations

Others have conducted studies of early scientific prose by a close reading of articles and books by historically significant scientists like Boyle,[10] Newton,[11]

9 Joseph Ben-David, *The Scientist's Role in Society. A Comparative Study* (Englewood Cliffs, NJ: Prentice Hall, 1971), 82.

10 Steven Shapin, "Pump and Circumstance. Robert Boyle's Literary Technology," *Social Studies of Science* 14, 4 (November 1984), 481–520; Jan V. Golinski, "Robert Boyle. Scepticism and Authority in Seventeenth-Century Chemical Discourse," in *The Figural and the Literal. Problems of Language in the History of Science and Philosophy, 1630–1800* eds. Andrew E. Benjamin, Geoffrey N. Cantor, and John R.R. Christie (Manchester: Manchester University Press, 1987), 58–82.

11 Alan G. Gross, *The Rhetoric of Science* (Cambridge, MA: Harvard University Press, second edition, 1996), 119–128.

and von Guericke,[12] or by an analysis of a selection of *Philosophical Transactions* articles.[13] To our knowledge, however, none has attempted systematically to compare the communicative practices of the two centres of 17th-century science, England and France, using a sample of broad scope.[14] Our object of study for this purpose is the scientific journal article, a genre of discourse that originated independently in England and France during the year 1665.

In our view, one of the methodological weaknesses of past rhetorically-informed analyses of scientific texts has been: how does one know whether the insights into scientific communication represent anything more than the possibly idiosyncratic characteristics of particular authors? Moreover, how does one know whether emphasis on a distinctive rhetorical feature or two doesn't leave the reader with a distorted impression of the scientific literature as a whole? For example, much has been made of the epistolary strains of the early articles in the *Philosophical Transactions,* as manifested by the occasional opening of "Dear Sir" and the rare tidbits of extraneous personal information. Yet our own impression is that the typical scientific articles in other early scientific journals little resemble what one normally thinks of as a "letter."[15]

As a first step toward overcoming such limitations, we have attempted what we hope will prove to be a more systematic approach to a rhetorically informed analysis. Our approach has three main components: guided by past studies on

---

12 Charles Bazerman, "Forums of Validation and Forms of Knowledge. The Magical Rhetoric of Otto von Guericke's Sulfur Globe," *Configurations* 1 (1993), 201–227.

13 In addition to the work of Allen, Qin and Lancaster, discussed in note 3, see: Charles Bazerman, *Shaping Written Knowledge. The Genre and Activity of the Experimental Article in Science* (Madison: University of Wisconsin Press, 1988), 59–79; Ellen Valle, "A Scientific Community and Its Texts. A Historical Discourse Study," in *The Construction of Professional Discourse*, eds. Britt-Louise Gunnarsson, Per Linell, and Bengt Norberg (London: Longman, 1997), 76–97; Dwight Atkinson, *Scientific Discourse in Sociohistorical Context. The Philosophical Transactions of the Royal Society of London, 1675–1975* (Mahwah, NJ: Lawrence Erlbaum Associates, 1999). Our study differs from these earlier ones in several respects, most importantly in the comparison with the French journals, but also in our method for selecting and analyzing texts.

14 For a historically-based analysis of select rhetorical features of English prose in the 17th century, see Peter Dear, "*Totius in verba.* Rhetoric and Authority in the Early Royal Society," *Isis* 76, 282 (June 1985), 145–161. For a historical treatment of early French scientific articles, see Frederic L. Holmes, "Argument and Narrative in Scientific Writing," in *The Literary Structure of Scientific Argument,* ed. Dear (Philadelphia: University of Pennsylvania Press, 1989), 164–181.

15 In *De Elocutione,* Demetrius noted: "If anybody should write of logical subtleties or questions of natural history in a letter, he writes indeed, but not a letter"; quoted in Abraham J. Malherbe, *Ancient Epistolary Theorists* (Atlanta, GA: Scholars Press, 1988), 19.

the rhetoric and history of science, we start by assembling a well-motivated set of questions relevant to scientific argument in the 17th century; then we ask these questions of a large, representative sampling of texts from three journals; finally, we tabulate and interpret the results, and infer from them our explanation of their possible sociological implications.

Consistent with the spirit of Chaïm Perelman and Lucie Olbrechts-Tyteca's *The New Rhetoric*,[16] we define argument broadly as the ensemble of means that authors employ to make and support their assertions. In particular, we concentrate on how scientists employ argument to establish new facts about the material universe and offer explanations for them. We propose four topics as central to understanding scientific argument in the 17th century. These topics are based on what we take to be a consensus among scholars who have dealt with scientific communication in the 17th century.[17] While other taxonomies are possible, we take ours as at least plausible:[18]

1. What makes an observation or an experimental result a scientific fact? In the 17th century, there were at least two standards according to which facts were judged scientific. First, a scientific fact was an occurrence in the natural world that was reported by a reliable individual—usually a member of a scientific society—and was witnessed by one or more such observers. Second, a scientific fact was a finding resulting from the appropriate use of instruments (telescope, microscope, thermometer, air-pump, and the like) combined with methods discussed in item 2.
2. How are the facts generated? Whether the 17th or 20th century, there are several basic means: observation of objects and events in natural settings; manipulation of natural and man-made objects in artificial settings for experimental purposes; computation or description based on theory; or some combination of these. The fact-generating enterprise of science also entails developing new methods and equipment for improving

16 Chaïm Perelman and Lucie Olbrechts-Tyteca, trans. John Wilkinson and Purcell Weaver, *The New Rhetoric. A Treatise on Argumentation* (Notre Dame, IN: University of Notre Dame Press, 1969).

17 See Shapin, *op. cit.* note 10; Gross, *op. cit.* note 11; Bazerman, *op. cit.* note 13; Dear (1985), *op. cit.* note 14; Holmes, *op. cit.* note 14; Peter Dear, *Discipline and Experience. The Mathematical Way in the Scientific Revolution* (Chicago, IL: The University of Chicago Press, 1995); Lorraine Daston and Katharine Park, "Strange Facts," in their *Wonders and the Order of Nature, 1150–1750* (New York: Zone Books, 1998), 215–254.

18 In the Appendix at the end of this paper, we present the set of questions we derived from the topics of argument.

observations, conducting experiments, and confirming or extending theoretically based assertions.

3. What is the relationship between scientific facts and scientific explanations? In the 17th century, not all scientists were concerned with explanation; many were content with collecting and ordering facts about nature. Scientists interested in explanation, however, realized that their explanatory conjectures must square reasonably well with the facts defined according to item 1.
4. What sorts of explanation are there? For 17th-century scientists there were two kinds of explanation, both causal: mechanical and mathematical. Sometimes mechanical causes were directly observable. Sometimes they were observable only by their effects. In that case, plausible reconstruction was necessary. But causes might also take the mathematical form of universal laws. These would have a mathematical expression, consistent with the Galilean programme: $f = ma$, for example.

In the remainder of this paper we will illustrate these topics and the range of arguments they generate. We consider not only written arguments but visual contributions to argument as well. As part of this discussion, we also examine how 17th-century scientists employed textual features known to be of argumentative importance to the modern scientific article: "hedging,"[19] and quantitative statements.[20]

To gather a representative sample for analysis, we randomly selected 100 whole articles from the three most significant journals reporting 17th-century science, as determined by Robert Gascoigne: *Philosophical Transactions, Journal des Sçavans* and *Mémoires de l'Académie Royale des Sciences.*[21] For the analysis of hedging and quantitative statements, we gathered ten-line passages

19 Ken Hyland, *Hedging in Scientific Research Articles* (Amsterdam: John Benjamins, 1998); Greg Myers, "The Pragmatics of Politeness in Scientific Articles," *Applied Linguistics* 10 (1989), 1–35, esp. 12–20.

20 Theodore M. Porter, *Trust in Numbers. The Pursuit of Objectivity in Science and Public Life* (Princeton, NJ: Princeton University Press, 1995), 73–86.

21 Robert Mortimer Gascoigne, *A Historical Catalogue of Scientific Periodicals, 1665–1900, with a Survey of Their Development* (New York: Garland Publishing, 1985). In this book, Gascoigne ranks the available 17th- and 18th-century journals on the basis of the number of prominent "scientists" who published in a given journal over a select period. The other two 17th-century journals listed by Gascoigne, *Miscellanea curiosa* and *Acta eruditorum,* originated in German-speaking lands and adopted Latin to reach the widest possible audience.

randomly selected from 200 articles in the same three journals. We divided the articles and passages equally among the English and French.

We anticipated that tabulating the answers to our questions would provide us with an empirical foundation upon which to erect plausible generalizations about the argumentative practices of the English and French scientific communities in the process of formation. We make no pretence of having isolated all the argumentative features of the 17th-century scientific article, or of having entirely avoided arbitrariness in our questions, or of avoiding some subjectivity in our application of them. We are confident, nevertheless, that our method yields conclusions sound enough for confirmation or refutation about the general state of scientific argument in the various scientific disciplines, and in the English and French communities out of which modern science emerged. (For ease of reading, we have translated all the French passages quoted below into English.)

Arguing the Facts into Place

In reading the English and French articles in our sample, one is immediately struck by the *miscellanea curiosa* of new facts about the natural world and its workings. Monsieur de la Voye reports on little worms that eat stone (1666-MAS010; also in 1676-PT082).[22] Lister speculates on the existence of veins in all kinds of plants (1672-PT063). Méry describes the physiology of a pelican (1693-MAS084). Homberg reports on why leaf-like crystals formed in a flask full of antimony and rainwater (1693-MAS075), and also on how he was able to distil phosphorus from urine (1692-MAS009). Halley calculates the life expectancy in the city of Breslaw (1693-PT041). Henshaw offers "some observations and experiments upon may-dew" (1665-PT102). La Hire witnesses and measures the progress of an eclipse of the moon on 28 July 1692 (1692-MAS045). Hooke provides instruction on how to "make the picture of any thing appear on a wall, cub-board or within a picture-frame" (1668-PT006). Cassini observes through his telescope "a precise conjunction of a satellite of the planet Saturn with a fixed star" (1692-MAS074).

The data for genre in Table 4.1 indicate that observation is by far the favourite means of establishing new facts in our 100-article sample. About half this sample presents observational facts exclusively, or observational facts

22 These references indicate year of publication, journal abbreviation (PT = *Philosophical Transactions*, JS = *Journal des Sçavans*, and MAS = *Mémoires de l'Académie Royale des Sciences*), and our working code number.

TABLE 4.1 *Genre and purpose in sample of 100 whole articles*

	French		English	
	No.[a]	%	No.[a]	%
Genre				
Experimental	5	10	6	11
Theoretical	6	12	10	19
Methodological	3	6	5	9
Observational	10	20	20	38
Observational/theoretical	18	35	7	13
Experimental/theoretical	4	8	1	2
Mathematical	5	10	3	6
Other	0	0	1	2
Main Purpose				
Observation	11	21	19	36
Experimental results	5	10	8	15
Mechanical explanation	19	36	14	27
Mathematical rule	6	12	2	4
Mathematical explanation	10	19	3	6
Making or improving equipment	1	2	2	4
Other	2	4	4	8

[a] As the categories are not completely exclusive, totals add up to more than 50.

buttressed by some theoretical discussion. Also noteworthy in Table 4.1 is that of the 71 observational and experimental articles, more than half offer new facts without any theoretical discussion. We know from Steven Shapin and Simon Schaffer that English authors from the 17th century, particularly Robert Boyle, "offered the matter of fact as the foundation of proper knowledge."[23] The animating philosophy behind this approach is that theories come and go and are constantly being revised and expanded; facts "mirror nature." For that reason, early researchers like Boyle offered theoretical explanations of their observational and experimental "facts" only with extreme caution.

23 Steven Shapin and Simon Schaffer, *Leviathan and the Air-Pump. Hobbes, Boyle, and the Experimental Life* (Princeton, NJ: Princeton University Press, 1985), 24.

This orientation toward observed particulars, however, does not mean that theory is anathema in the 17th-century article. The data in Table 4.1 show that 54 of the 100 articles possess some theoretical or mathematical discussion. This interest in the abstract and conceptual appears to be greater in the French than in the English. True, in our sample, the English were as likely as the French to compose purely theoretical or mathematical articles (13 and 11 articles, respectively). Nevertheless, in the English sample, only 8 articles moved from observation or experiment to theory, while in the French sample, 22 articles did so.

Whatever the means of establishing facts, English and French scientists are both concerned about the warrant for them. Clearly, they must meet a higher standard than the opinions and observations of everyday life, if only because authors know that their statements will likely receive scrutiny and judgement by an international group of readers with similar interests.[24] In deference to this standard, we know from Myers and others that modern scientists hedge.[25] A simple textual feature, barely noticeable to the uninitiated, "hedging" is their indication that the quality of evidence is being carefully weighed. This feature is present but not prominent in 17th-century scientific prose. As evidence for this claim, we found relatively few hedges in our 200 short passages: averages of 0.13 per hundred words for the French, 0.21 per hundred words for the English,[26] an appropriately small amount for narratives relying heavily upon personal observations as well as mathematical calculation.[27] Typical are assertive

24 This expectation did not, of course, eliminate the occasional appearance of unfounded or ludicrous assertions. In his *Review of the Works of the Royal Society* (London, 1751), John Hill mercilessly criticized *Philosophical Transactions* for publishing "trivial and downright foolish articles" over the previous century. As evidence, he cited articles on a merman discovered in Virginia, and a miraculous plant that heals fresh wounds ("but to touch it, is to be healed"), as well as "incontestible proofs of a strange and surprising Fact, namely, that Fish will live in Water." Not long after Hill's scathing attack, in 1752, the Society established a committee to review articles for publication on the basis of "the importance and singularity of the subjects, or the advantageous manner of treating them": see Stimson, *op. cit.* note 4, 141. Worth noting, however, is that our sample of 100 randomly selected English and French articles did not have any obvious examples of ill-advised articles like those criticized by Hill.

25 Myers (1989), *op. cit.* note 19, 12–20.

26 The much lower French average for hedging may reflect inherent differences in the two languages, as well as less concern with the open display of gentlemanly politeness than among the English.

27 For comparison, we found that an analogous sample drawn from the 20th century yielded hedge counts of 1.82 and 1.29 instances per 100 words for English and French scientific articles, respectively.

sentences without qualifications, such as: "The species of corals and of *Madrepora,* the mushrooms of the sea, the *Tubularia marina, rubra IB* are also as hard as stones" (1700-MAS040), and "In this equation no more is needed than the substitution of the values of *v* & of *dz,* which will result in *x* & in *dx* of the equations of the given curves *AHG* & *AFK;* & will become that of the curve we are looking for, *BC*" (1699-MAS067).

Despite the low counts for hedging, we found many interesting varieties of 17th-century scientific hedging populating select passages from our sample. In this heavily hedged passage from *Philosophical Transactions* on the existence of veins in all kinds of plants, for example, Martin Lister leaves no doubt that he is speculating about the ramifications of his collected facts (hedges in bold):

> That **though we seem to be more certain** of the ramifications of the Fibres, wherein those veins are, **we yet are not so**, that those veins do any where grow less and smaller, **though probably it may be so**. That which makes us **doubt** it, is the exceeding smallness of these veins already, even where we might **probably expect** them to be Trunk veins and of the largest size; and being there also in very great Numbers and running in direct lines along the fibre, we **guess**, that one or more of them **may be** distributed and fall off on either hand with the subdivisions of the fibres, and not suffer any diminution in their bulk.
>
> 1672-PT063

In a French passage, Méry uses hedges to dramatize the tension the scientist feels between the doubt he must always entertain and the relative certainty he can occasionally attain and enunciate:

> One **cannot doubt** that the small muscles that are attached to the feathers of the skin of the body of the pelican serve to move them in different directions, & that when they move, one after the other, they can make the feathers move in circles. It **certainly looks as if** the fleshy fibres of the down can also create the same movements.
>
> Mr Méry was not, in the pelican he dissected in 1686, looking for the place from which the air came that filled the cells of the skin, but in 1692 he dissected yet another, in which he viewed it in a manner that **fully satisfied him.**
>
> 1693-MAS084

In the French sample, unlike the English, doubt has been expanded to include a set of concerns that focus on instrumental limitations, including that oldest of instruments, the human eye. While Méry's passage on animal physiology parallels Lister's, there is no English parallel in our sample for the following passage of Cassini's, which makes the limitations of nakedeye observations its subject. Cassini's is a second-order scepticism not found in the English sample. Cassini does not merely use hedges; he reflects on hedging:

> Scarcely does one find four or five observations of these conjunctions among all those that have been preserved since the beginnings of astronomy up to the present century: even so there is room to doubt whether these four or five apparent conjunctions, **since they were observed only with the naked eye**, are really precise and without any interval. For we now know that, because of light rays that increase the apparent size of the stars, **there are conjunctions that appear to be precise, although they may be nothing of the sort**; the use of telescopes having forced us to recognize that there are considerable intervals between stars that appear to the naked eye to be so completely conjoint that they seem one and the same star.
> 1692-MAS080

For English scientists, the warrant for facts seems more a matter of reliable testimony than of precise measurement. In the following English passage on a medical curiosity, scientific evidence is viewed exactly as forensic evidence would be, the absence of hedges being implicitly justified by the presence of sworn testimony:

> We, the Physicians under-named, do hereby certify, that all the particulars of the Chirurgical operation of extracting the Bodkin out of the Bladder of *Dorcas Blake,* as contained in the foregoing account, are truly and faithfully related, it being perform'd by Mr *Thomas Proby,* Master Chirurgeon, with great skill and success in our presence, as witness our Hands, this 22d of *May,* 1695.
> *I. Madden* Presid. *T. Molyneux.*
> *Wm. Smith.*
>
> *Dorcas Blake,* of *Fishamble-street,* in the Parish of *St John* [,] *Spinster,* came this day before me, and being sworn on the holy Evangelists, saith that the above relation is true in substance, and that she did swallow the Bodkin therein mentioned, and that the Bodkin now shewed to her is the same that she formerly swallow'd.
>
> her
> *Dorcas* X *Blake*
> mark.

Jurat coram me decimo die Junii 1695.

G. Blackall *Major. Dublin.*
(It is sworn in my presence, 10 June, 1695
G. Blackall, Sr., Dublin)
1700-PT110

In 17th-century science, it was at least an occasional practice to lend credibility to reported facts by mentioning witnesses. In our sample of 100 whole articles, this practice occurs in articles reporting some unusual observation or especially important experimental results, and appears to be more characteristic of the English than the French. In the English sample there are five articles citing witnesses: two experiments with animals (1667-PT003 and 1696-PT014), one physics experiment by Boyle (1676-PT021), one trial at Court involving some medication for coagulating blood (1673-PT034), and one monstrous birth (1670-PT038). Only two French articles mention witnesses: a joint observation by a group of astronomers (1677-JS042) and an experimental series with an English connection (1672-JS006)—namely, the article had been read at a Royal Society meeting. This stronger reliance on personal testimony is indicated in English scientific prose through an orientation toward human relationships largely absent in the French: witnesses are sworn, observers of experiments are named, readers are invited to repeat experiments. An orientation toward readers is also embodied in expressions that "humanize" the prose to the detriment of its objective stance. When a Fellow of the Royal Society says, "I shall premise some Experiments made many years ago which perhaps may give no less satisfaction unto many of our Experimental Philosophers than they afforded me, when I first made mine" (1674-PT072), he is deferring to his colleagues' own past experiences with conducting and witnessing experiments, not expressing doubt.

We depart here briefly from our random sample to address an argument topic of 20th-century science that seems to be largely absent from its 17th-century ancestor: replicability as a check on experimental or observational quality and reliability. We do not mean that experiments and observations are not repeated or contested. In the case of the controversy surrounding Newton's first scientific article (1672)—reporting the optics experiments he carried out in his darkened chamber with a glass prism—the durability of purported facts as indexed by the repetition of experiments is clearly an important topic of argument.[28] But while Robert Hooke and Christiaan Huygens challenged

28 This controversy, the first major open dispute in the scientific journal literature, has been much analyzed by both historians and rhetoricians of science. See, for example, Gross,

Newton's theoretical interpretation of his experimental results, neither questioned the results themselves. It was not until the letters from Francis Hall (also known as Linus, a Jesuit professor at the English College in Liège) arrived in 1674, that the experiments were even performed at the Royal Society, the "house of experiments." Even in this case, the issue was less the durability of purported facts than the official witnessing of Newton's experiments by other Society Fellows. In the *Opticks* of 1704 (a work just past our period), Newton, in enjoining his readers to repeat experiments, is encouraging them, not to check, but to see for themselves. The case of Huygens' replication of some of Boyle's experiments with an air-pump, recounted by Shapin and Schaffer, is closer to the modern standard: but it seems to be a notable exception to the rather casual rule; as Shapin and Schaffer note,

> ...even the notion of *verification* itself is profoundly problematic. The Florentines [Accademia del Cimento in Italy] announced that they had verified Boyle's results without needing Boyle's machine.[29]

Historian Alan Shapiro summarizes what we take to be the most accurate view:

> The very concept of rigorous and public replication, which is often considered to play a central role in modern science...was by no means a standard feature of seventeenth and early eighteenth century science.... [T]he experimental tests of Newton's theory were private, casual, and, by later standards, rather lax in what they took to be either confirmation or refutation.[30]

Arguing Explanations into Place

Our English and French samples differ not only in their manner of establishing new facts but also in their interest in explaining these facts. In our sample of English articles, 20 made observational claims detached from any apparent explanation and 6, experimental claims detached from any apparent explanation. In the French sample, 10 articles made only observational claims and 5, only experimental claims. So we tallied 26 factual claims independent

op. cit. note 11, 111–128; Bazerman, *op. cit.* note 13, 80–127; and Richard S. Westfall, *The Life of Isaac Newton* (Cambridge: Cambridge University Press, 1993), 85–109.

29 Shapin & Schaffer, *op. cit.* note 23, 225–282, quote at 281.

30 Alan Shapiro, "The Gradual Acceptance of Newton's Theory of Light and Color, 1672–1727," *Perspectives on Science* 4 (1996), 59–140, at 61.

of causal explanation for the English, but only 15 for the French sample (see Table 4.1).

These bare numbers hint at the differing scientific practices and concerns in each country, but they do not give a firm sense of the flavour of two sorts of 17th-century scientific argument, one presenting new facts, the other providing an explanation for them. To illustrate this difference, we selected two articles: Thomas Henshaw's *"Some Observations and Experiments upon* May-Dew" and "Extract from a Letter of Mr. Huygens, touching the Phenomena of Water from which all Air has been Removed." In the first case, Henshaw subjects May-dew (dew gathered in May and supposed to have medicinal and cosmetic properties) to a series of trials designed, apparently, to ascertain the conditions under which it will putrefy. We say "apparently" because this general purpose is nowhere stated; it must be inferred from a series of loosely connected experiments narrated in no discernible order. The series leads to no conclusion and is preceded by a narrative introduction, not about Henshaw's experiments, but about Henshaw himself, presumably written by the editor of *Philosophical Transactions:*

> That ingenious and inquisitive Gentleman, Master *Thomas Henshaw,* having had occasion to make use of a great quantity of *May-dew,* did, by several casual Essayes on that Subject, make the following Observations and Tryals, and present them to the *Royal Society.*
>
> 1665-PT102

Henshaw's experiments are reported in a lively, novelistic manner, a mode of representation that leaves their larger purpose unclear. Here's a single sentence vividly weaving together method and results, but offering little in the way of explanation:

> That having several Tubs with good quantity of *Dew* in them, set to putrefy in the manner abovesaid, and comming to pour out of one of them to make use of it, He found in the water a great bunch, bigger than his fist, of those Insects commonly called *Hog-lice* or *Millepedes,* tangled together by their long tailes, one of which came out of every one of their bodies, about the bigness of a Horsehair: The Insects did all live and move after they were taken out.
>
> 1665-PT102

In contrast, Huygens' focus is explanation. His *Journal des Sçavans* article has three main components: a theoretical introduction that motivates his investigation; a series of experiments designed specifically to answer the theoretical question he poses; and a conclusion that ponders the theoretical impact of his results. His introduction follows the essentials of 20th-century practice.[31] He stakes out an intellectual territory—one of Boyle's research programmes—and establishes his niche within it:

> Before communicating to you what I have observed concerning the suspension of water in a vacuum, I want to repeat the experiments, in order to verify the remarks I made at another time and in order to attempt to penetrate to the causes of an effect so surprising. I will first relate my observations and then I will pass to the conjectures I have made to account for them.
>
> The experiments that the illustrious Mr. Boyle performed one day in 1661, with the description of the pneumatic pump, gave me a starting point for examining the matter at hand. In one of these experiments, Boyle put a four-foot glass tube, full of water, in a container or vessel from which the air had been evacuated, the open end of the tube being submerged in another quantity of water, contained in a glass. The air from the container having been evacuated as far as was possible by means of his pump, the water of the tube flowed down into the glass up to the point at which no more than about one foot remained, all the rest of the tube remaining empty of water and of air. He judged with good reason that this height of one foot of water that stayed above the level of that in which the open end of the tube submerged, remained suspended because there remained in the container a bit of air that the pump, being defective in that respect, could not evacuate.
>
> 1672-JS006

In the article itself, Huygens occupies the niche his criticism of Boyle creates, providing the reader with a new theory to explain his novel results. In so doing, Huygens challenges Boyle's reasonable inference by completing and reporting on a series of six experiments, each carefully described. Having depicted his apparatus, he launches into a narration about the trustworthiness of his method, a narration that is also a part of his argument:

> But at the end of the month of December in the same year, 1661, having left this water in the vacuum for twenty-four hours (which completely purged it of the bubbles it gives out when fresh water is used) & having

31 John Swales, *Genre Analysis. English in Academic and Research Settings* (Cambridge: Cambridge University Press, 1990), 137–166.

> filled the vial *C*, I was surprised to see that notwithstanding that I had tried very hard to remove the air from vessel *B*, the water did not descend at all from *C*, which remained perfectly full; I could hardly suspect that there had been any defect in my pump, nor that the vessel *B* was badly blocked; but to be perfectly clear about this, I removed the vial *C* from underneath the vessel and after having forced in a small air bubble, I replaced it as before; and having worked the pump, I saw that in the end all the water descended very nearly up to the level of that of glass *D*. That assured me that there had been no defect in the pump, and that the water purged of air remained suspended without descending, although the vessel *B* was completely void of air, or at least as much of it as it was when the fresh water descended from the vial. I made the water descend a second time by inserting into the neck of the vial a bubble so small that it was hardly visible.
>
> 1672-JS006

After describing in detail the six interlocking experiments, Huygens concludes with his new claim (the existence of a "matter…more subtle than air") and a recommendation for future work to extend his claim:

> Here is another confirmation of our hypothesis of a matter capable of exerting pressure but more subtle than air. Those who would take the trouble to discover up to what point the force of this pressure mounts can do no better than to pursue the experiment with the tubes full of mercury, even longer than those Mr. Boyle used. They will find perhaps that this force is powerful enough to cause the union of the parts of glass and of other sorts of bodies that hold together better than if they were joined only by contiguity and rest, as Mr. Descartes would have it.
>
> 1672-JS006

Huygens and Henshaw differ in their goals: only Huygens is interested in explaining the phenomena that experiment reveals. It is significant that the explicit view of the far better-known Robert Boyle is coincident with Henshaw's implicit one: Boyle insisted that his "business" was "not…to assign the adequate cause of the spring of the air, but only to manifest, that the air hath a spring, and to relate some of its effects."[32] Such a view contrasts with that of Huygens,

32 Steven Shapin, *A Social History of Truth. Civility and Science in Seventeenth-Century England* (Chicago, IL: The University of Chicago Press), 335. This is not to say, of course,

who designs a series of experiments whose purpose is to test a new hypothesis derived from a re-analysis of Boyle's experimental design.

Because Henshaw makes no claim other than that of fact, he needs no argument apart from the array of relevant data. He describes nature; he makes no attempt to understand its workings. Huygens, who makes a claim about those workings, needs a more robust argument. In satisfying this need, Huygens anticipates the form of the modern scientific article. This form is also an argument: it is meant to recapitulate, not the experimenter's story, but that of the experiment, a narrative that is also a logical progression. This progression justifies a conclusion that contains referents specific to nature, in Huygens' case, not to his elaborate tangle of glassware, but to "matter capable of exerting pressure but more subtle than air." The way in which he structures this argument is clearly consistent with Holmes' assertion that "the 'conventional form' imposed on the modern journal article is...but the outcome of the long evolution of a form that emerged during the late seventeenth and early eighteenth centuries,"[33] the main impetus coming from France under the auspices of the Académie Royale.

Our results in Table 4.1 (bottom portion) suggest that Huygens' concern for explaining his gathered facts is more characteristic of the French than the English scientific literature: 29 French articles offer explanations, while this is true of only 17 English articles. None the less, both the French and English place a healthy emphasis on the search for mechanical causes: of our 50 French articles, 19 offer mechanical explanation; of 50 English articles, 14 offer the like.

Both French and English explanations are, typically, plausible reconstructions of mechanical processes. On occasion, such mechanical explanations involve entities not visible to the explaining scientist: in Huygens' view, for example, the action of ordinary air is insufficient to explain the behaviour of devices like the barometer in the void created by an air-pump. To remedy this deficiency, Huygens hypothesizes "subtle air." Surprising as this may seem, in no English case in our sample of 50 articles does a mechanical explanation depend on speculations like that of Huygens concerning the basic constitution of the physical universe.

A more Galilean approach to explanation also seems to be reflected in our French sample: 16 articles have a mathematical purpose, as compared to only 5 English articles. This is consistent with Dear's assertion that "mathematical

---

that Boyle did not have in mind a causal explanation for the "spring of the air," only that he hesitated to spell it out for public consumption in his written communications.

33 Holmes, *op. cit.* note 14, 165.

contributions…fitted only uneasily into the Royal Society's [early] work."[34] A more mathematical approach by the French may likewise account for the differences in quantitative expressions and equations found in our random sample of 200 short passages. Our average counts per 100 words for this feature were 2.48 for the French versus 1.09 for the English.[35] This contrast must not be generalized too far, however: Newton, Wallis and Halley were as Galilean as Huygens or any French researcher.

This difference in emphasis on mathematics and the quantitative is best illustrated with examples. The first, culled from an early issue of the *Philosophical Transactions,* reveals the more personal, less mathematical bent of the English sample (quantitative expressions in bold):

> I took out **45 *ounces*** and better, of blood, out of the *Jugular Vein* of a Sheep, of a lesser size than the former; by which time, the Spectators, as well as my self, found her exceeding faint, and some thought her pass'd recovery, without a supply of blood. Then I convey'd blood from the *Jugular Vein* of a Calf into that of the Sheep for the space of ***7 minutes,*** when we did believe, by the continuance of a good stream from the Calf, that the Sheep had already received more blood, than she had lost. Whereupon we set her free, and she had no sooner got her liberty, but seeing a Dog near her (which was a Spaniel, that had formerly suffered the transmission of Sheeps-blood into him) she butted with great violence at him **three or four times,** not appearing at all concern'd at what she had endured in the Experiment.
>
> 1667-PT003

In this selection on the then-popular topic of experiments with blood transfusion, the author relates a series of events in the first person singular and plural. The three quantitative expressions act more as stage scenery than part of a coherent mathematical or mechanical explanation. Unlike 20th-century scientific prose, narrative particularity still seems to have more probative value than any measurements. The vivid scene in this passage, including the mention of "Spectators" and the seemingly irrelevant yet amusing aside delivered in the parenthetical remark, attests to that event having happened in just that way.

---

34 Dear, *op. cit.* note 14, 159.

35 For comparison, we found that a similar analysis of passages from the 20th century yielded 3.28 and 2.82 instances per 100 words for English and French scientific articles, respectively.

The text thus conveys the impression of the author as eyewitness to "an actual, discrete event," and an active participant in that event.[36]

The original of our next example is in French, and from the end of our chosen period. It concerns the making of water-clocks, and exhibits an orientation that is definitely quantitative and mathematical, as well as impersonal:

> For there to be these **144 equal parts**, it is necessary to have **23** for the **first hour** of outflow, **21** for the **second, 19** for the **third, 17** for the **fourth**, and, finally, **1** for the **twelfth**, always following the natural sequence of odd numbers.
>
> EXAMPLE II
>
> XI.* The acceleration of the water at its exit through the hole *O* remains the same as in the last example (*art.* 8), that is to say, as the roots of the heights of the surface of the water above the hole, so that the curve *OVX* that expresses them by these ordinates is still a parabola whose locus is $v = \sqrt{px}$; if one now wants the generated curve *FEO* of the given vessel, assume any conic section whose vertex is within the circumference of *O*.
>
> 1699-MAS023[37]

This passage has no proper names or first-person pronouns or any language referring to things extraneous to the science being reported. There is no sense of a particular place or an observer's presence, as in the first quoted passage, and the instances of quantitative expression all work in service of the author's mathematical explanation. The French author is clearly addressing colleagues, fellow labourers in the manufacture of new scientific knowledge—not the universal audience of "Artizans, Countrymen, and Merchants" imagined by Thomas Sprat in his famous pronouncement about the preferred style by Royal

36 Dear, *op. cit.* note 14, 152. Note that our sample also contains many passages focused on the natural world, with the observer more in the background, as in the earlier quoted passage of Lister (1672-PT063). Furthermore, in our sample of 100 English passages, we found little evidence that "Fellows of the Royal Society *filled their descriptions* with contingent details of time and place in order to convey the impression that these were actual events, of which the authors were faithful reporters" (Holmes, *op. cit.* note 14, 165, italics added). In the midst of maritime observations, the Englishman Stubbs does report that his ship sailed off course "to avoid the *Spanish* Fleet" (1668-PT010), but only a few such apparently extraneous statements appear in our sample. Our own general impression is that when there are details of time and place, as is often the case in reports of astronomical observations, they almost always provide essential information to the technical matter at hand.

37 The original French passage reads as follows:

Society members.[38] Still, the high standard deviation for quantitative expressions and equations in the French sample (±7.24 *versus* ±2.73 for the English) suggests that such mathematical text as the quoted French passage occurs in a relatively small percentage of samples. Overall, in the 17th century, the data within scientific articles are relatively meagre and their descriptive complexity not very far advanced, even in so mature a science as astronomy.

Depicting Argument

Visual representations also contribute to the argumentative practices at work in our 17th-century scientific texts. Because during this period the technology for reproducing figures (woodcuts, copper engravings and etchings) was still in its infancy and expensive,[39] it is not surprising to find that, of our 100 whole articles (50 French, 50 English), only 40 possess any visual representations at all. Of these 40, the average space taken up by the visuals in each article is fairly modest: 10% in the French case, 8.3% in the English.[40] These visuals vary in type—tables of results, three-dimensional drawings of equipment and of objects being investigated, and geometric

---

De sorte que de ces 144 parties égales il en faut prendre 23, pour le premiere heure de l'écoulement; 21, pour la seconde; 19, pour la troisiéme; 17, pour la quatriéme; & enfin 1, pour la douziéme, toûjours suivant la suite naturelle des nombres impairs.

EXEMPLE II

XI.* La vîtesse de l'eau à la sortie par le trou *O*, demeurant la même que dans l'exemple précedent (*art.* 8), c'est à dire, comme les racines des hauteurs de la surface de l'eau par dessus ce trou, ensorte que la courbe *OVX* qui les exprime par ses ordonnées, soit encore une parabola dont le lieu soit $v = \sqrt{px}$; si l'on veut présentement que la courbe géneratrice *FEO* du vase donné, soit une section conique quelconque dont le sommet soit en *O*.

38 Thomas Sprat, *History of the Royal Society of London, for the Improving of Natural Knowledge* (London: J. Martyn & J. Allestry, 1667), 113.

39 The expense did not impede the production of books that were magnificently illustrated and designed, such as Robert Hooke's *Micrographia. Or, Some Physiological Descriptions of Minute Bodies Made by Magnifying Glasses* (London, 1665) and the Accademia del Cimento's *Saggi di Naturali Esperienzi* (Florence, 1667). For a vivid portrait of the costs and perils associated with 17th-century publishing in natural philosophy, see Adrian Johns, *The Nature of the Book. Print and Knowledge in the Making* (Chicago, IL: The University of Chicago Press, 1998), 444–542.

40 Printing technology at this time allowed for text to accompany tables, but engravings are inserted separately (often at the end of the journal issue), and the relative space taken up

representations of the operation of natural laws.[41] More important, they vary in function. Function is where their real significance lies: they provide support for the facts and the mechanical and mathematical explanations that are the central topics of argument in 17th-century science (see Table 4.2). The following are two examples from our French sample of the ways in which 17th-century scientists employed visuals to establish facts and offer explanations for facts.

Huygens' illustration (Figure 4.1), taken from the article discussed previously, is a typical visual representation of an experimental apparatus designed to establish facts of nature through laboratory manipulations. Snugly embedded in the text, the stick-figure-like drawing makes use of dotted and solid lines, shading and labelling, and keeps extraneous details and embellishments to a minimum, so as better to represent the various experimental components within components (1672-JS006). This is not a deeply realistic rendering of an experimental apparatus, as in Boyle's exceptional engraving of his air-pump printed in *New Experiments Physicomechanicall, Touching the Spring of the Air, and Its Effects (Made, for the Most Part, in a New Pneumatical*

TABLE 4.2 *Functions of graphics in sample of 100 whole articles (percentage of total graphics given in parentheses)*

	French articles	English articles
Supports observation	8 (31)	6 (43)
Supports experimental result	2 (8)	1 (7)
Supports mechanical explanation	3 (11)	1 (7)
Supports mathematical relation	8 (31)	3 (21)
Supports mathematical explanation	2 (8)	2 (14)
Supports equipment building	3 (11)	1 (7)
Total graphics	26	14

by illustrations depends on how many fit on to a single plate, not on their argumentative importance to the articles they illustrate.

41 For a discussion of the later use of graphs, their rôle in argumentation, and the distribution of their use across 20th-century science, see Laurence D. Smith, Lisa A. Best, D. Alan Stubbs, John Johnston, and Andrea Bastiani Archibald, "Scientific Graphs and the Hierarchy of the Sciences. A Latourian Survey of Inscription Practices," *Social Studies of Science* 30, 1 (February 2000), 73–94, esp. 74–76, 84–87.

Engine),[42] but a simplified abstraction designed better to reveal internal workings of the apparatus.

Typical also is Huygens' careful explanation of this device in the running text of his article (in lieu of a figure caption), an investigation of the anomalous height of water in a tube under a vacuum created by an air-pump:

> It is necessary to imagine that the glass denoted *C C* is completely full of water, & that its open extremity dips in the water of glass *D*. Over the top of both is placed the vessel *B*, on whose open mouth is applied a certain soft cement, spread on the plate *A A*, which is pierced by a little hole in the middle, by which the air leaves when one works the pump. When I used fresh water, all of vessel *C* emptied until it was at the level with that of glass *D*.
> 1672-JS006

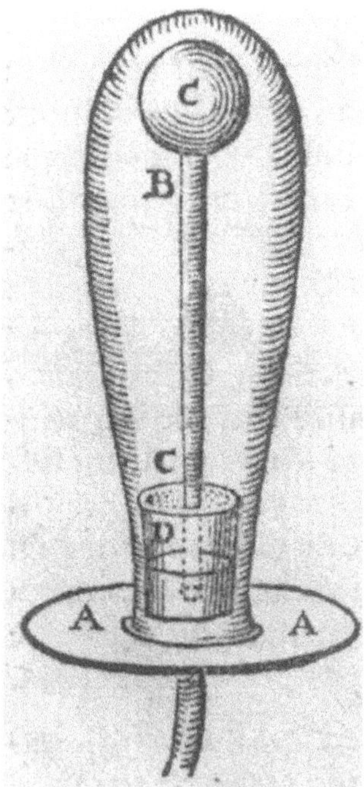

FIGURE 4.1 *Schematic of apparatus used in air-pump experiments*
SOURCE: FROM CHRISTIAAN HUYGENS, 'EXTRAIT D'UNE LETTRE DE M. HUYGENS DE L'ACADÉMIE ROYALE DES SCIENCES À L'AUTEUR DE CE JOURNAL, TOUCHANT LES PHÉNOMENES DE L'EAU PURGÉE D'AIR' [EXTRACT FROM A LETTER…CONCERNING THE SURPRISING BEHAVIOUR OF WATER FROM WHICH ALL AIR HAS BEEN REMOVED], *JOURNAL DES SÇAVANS* (1672), 112–123, ON 113.

42 See Shapin & Schaffer, note 23, 27.

Huygens is interested in describing the experiment so carefully that through his own words, or through subsequent replication, the fact of this anomalous liquid suspension can be established. The drawing itself gives us no clue, however, as to the mechanical cause for this fact.

Méry's illustration (Figure 4.2) of two foetuses enclosed in the same membrane is an example of a realistic drawing of an object of study designed to illustrate mechanical cause (1693-MAS013). The visual is clearly at the service of science: each of the technical features—the amnion, the chorion, the placenta and, especially, the knotting of the two umbilical cords—is made visually prominent by means of some combination of contrast, position and labelling. Moreover, the explanation of the phenomenon—the mechanical cause of the strangulation alluded to in the article—is made visually the most prominent item, placed at the very centre of the picture, and marked by the letter "N." The grotesque yet accurate rendering of the foetuses flanking the

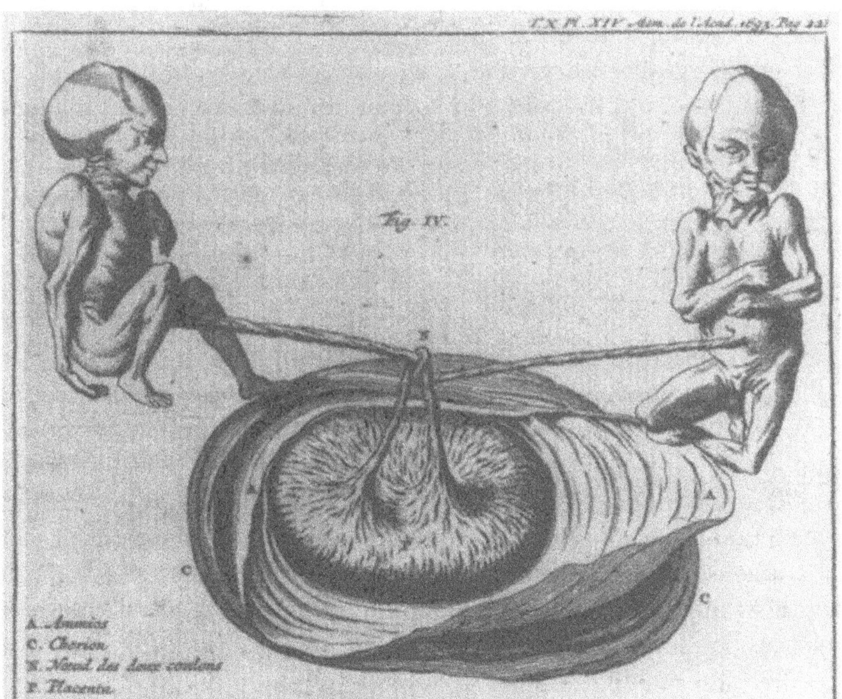

FIGURE 4.2 *Drawing of two strangled foetuses*
SOURCE: FROM JEAN MÉRY, 'OBSERVATION DE DEUX FOETUS ENFERMÉS DANS UNE MÊME ENVELOPPE' [OBSERVATION OF TWO FOETUSES ENCLOSED IN THE SAME CHORION], *MÉMOIRES DE L'ACADÉMIE ROYALE DES SCIENCES* 10 (1693), 201, ILLUSTRATION ON FOLLOWING PAGE.

knotted cord dramatizes the tragic consequences of this medical phenomenon (the wasting and ultimate death of the foetuses resulting from insufficient oxygen and nourishment through the cord), the human toll of which is outside the bounds of normal scientific discourse. This toll includes the demise of the nearly invisible mother, who is visually represented by her scientifically relevant anatomical feature alone.

While our sample of visuals for the 17th century is relatively small, we did find a few interesting differences between the French and English (Table 4.2). Whereas 26 (52%) of the French articles in the sample have tables or graphics, this is true of only 14 (28%) of the English articles, very possibly the result of the meagre financial backing of *Philosophical Transactions* by the Royal Society. More important, in the French and English cases, the distribution of the functions of the graphics differs somewhat. English visuals are more likely to support observational or experimental facts (50%, compared with 38% for the French articles with visuals), while French visuals are more likely to support a mechanical or mathematical explanation, or a mathematical relation (50%, compared with 43% for the English articles with visuals). Thus we find in the French case a greater use of visuals, particularly of a mathematical and explanatory nature, a use eventually to become a crucial element in scientific communication.

The Scope of Argument

We want to make a final point concerning, not argumentative practices *per se,* but their scope: English scientists exhibit a far greater breadth of subject matter than their Continental colleagues (Table 4.3). In the English case, in fact, we have one category of article for which there are no French instances: "miscellaneous."[43] The focus of the diverse articles in this category is generally on areas of human inquiry new to scientific investigation: interests are represented that would later become cartography, metallurgy, cultural anthropology and actuarial science. As a consequence of this broader English interest, the argumentative standards of 17th-century science are applied to a broader range of subjects. Science becomes more a way of looking at the world in general than

43 All of our categories of science are anachronistic, with the possible exceptions of astronomy, chemistry, mathematics and medicine. In applying these categories to the 17th century, we have attempted merely to identify what these men were doing, with no pretence that they would have identified their tasks exactly as such. Some of our other terminology must be so viewed: for example, "scientist," and even "experiment."

TABLE 4.3 *Distribution according to discipline for articles containing 200 selected passages*

	French	English
Animal Biology	16	16
Astronomy	21	12
Chemistry	11	5
Earth Sciences	5	6
Mathematics	10	7
Medicine	0	12
Physics	21	16
Plant Biology	13	7
Technology	3	2
Miscellaneous	0	17

a way of investigating the natural world, narrowly conceived. Halley's "An Estimate of the Degrees of the Mortality of Mankind, drawn from curious Tables of the Births and Funerals at the City of Breslaw; with an Attempt to ascertain the Price of Annuities upon Lives," is a good example of this extension and its communicative implications:

> For these Reasons the People of this City seem most proper for a *Standard;* and the rather, for that the *Births* do, a small matter, exceed the *Funerals.* The only thing wanting is the Number of the whole People, which in some measure I have endeavoured to supply by comparison of the *Mortality* of the People of all Ages, which I shall from the said Bills trace out with all the accuracy possible.
>
> It appears that in the Five Years mentioned, viz. from 87 to 91 inclusive, there were *born* 6193 Persons, and *buried* 5869; that is, born per *Annum* 1238, and *buried* 1174; whence an *Encrease* of the People may be argued of 64 per *Annum,* or of about a 20th part, which may perhaps be ballanced by the Levies for the *Emperor's* Service in his Wars.
>
> 1693-PT041

In this passage of actuarial findings, both Halley's style and his style of argument are indistinguishable from their counterparts in physics, biology and chemistry. In contrast, the professional status of French science is manifest in its sharper focus on subjects that are traditionally scientific. A group of scientists who clearly see themselves as astronomers, botanists, chemists, physicists or mathematicians carry out organized programmes of research.[44] Medicine is not among their chief concerns,[45] nor are they driven by a general desire to create new disciplinary interests by focusing their attention on objects and events hitherto excluded from scientific scrutiny.

Social Implications

In discussing the interesting differences between English and French scientific literature we have necessarily slighted the many similarities. Most articles in our sample could have been published, with some minor revisions and translation, in the other nation's journal literature; indeed, *Philosophical Transactions* did occasionally publish English and Latin versions of articles by European authors, including such prominent ones as Leeuwenhoek, Huygens, Hevelius, Leibniz, Varignon and Cassini. Overall, we see the writers of 17th-century scientific prose, whatever their nationality, as addressing an international community of readers engaged in a mutual quest for empirical knowledge of the natural world. Yet our rhetorically-informed analysis has revealed several different trends in argumentative practice that would appear to be national in character.

We have found that while 17th-century natural philosophers shared a common interest in fact-gathering and in explanation, the balance between these two enterprises differed from one shore of the Channel to the other. From words still vivid on the page, we can envision the English gentleman Thomas Henshaw, just after dawn on a chilly May morning, supervising his servants as they gather quantities of dew for his experiments. We cannot picture the mathematical Dutchman Christiaan Huygens similarly engaged. There is thus a special sense in which Henshaw participates in the great unrealized dream of English science, the dream in Bacon's *New Atlantis,* of a museum containing a specimen of every scientific fact. Although the flagship enterprise of Continental science—the Académie Royale—was not uninterested in Bacon's

44 Hahn, *op. cit.* note 4, 16.

45 This finding is a little deceptive in that some of the Académie Royale research in plant and animal biology had an underlying medical motivation. As Stroup points out (*op. cit.* note 4, 172–173), nearly 30% of the 17th-century Académie members had been trained as physicians or apothecaries, yet the Académie "established its scientific program to be independent of medical research." Also, at this time in France, medicine was already a well-developed discipline with various career tracks available, including academic ones.

dream, as its early work on animals and plants attests,[46] its focus on explanation was greater than that of the Royal Society. Furthermore, our French sample placed a heightened emphasis on quantification, visual argument, and use of observations and experimental results as stepping stones to theory.

The relationship between "literary inscriptions" and their sociohistorical context is never a simple one, and the reason behind the differences we have noted present a puzzle we cannot definitively solve. Nevertheless, we think that any explanation must involve the openness of the Royal Society versus the exclusivity that was a consequence of the professionalization of the Académie Royale.[47] Only members of the Académie Royale could publish in its *Mémoires,* a practice that skewed its contents in the direction of a small authorial community of academicians. *Journal des Sçavans* relied heavily upon this same limited community for its contents.[48] While the professionalization of science on a large scale would not take place until the middle of the 19th century, in Germany, when most working scientists were "either university teachers or students, and they worked more and more in groups consisting of a master and several disciples,"[49] the incipient professionalization of French science is already evident in the early years of the Académie Royale and the two journals in which its members published their research findings.[50]

In contrast, even though closely tied to the Royal Society, the early *Philosophical Transactions* published articles by men,[51] both inside and outside that community, including scientific correspondence sent to the editor from seafarers, physicians, world travellers, European savants, and the like. Thanks to its founder and first editor, Henry Oldenburg, *Philosophical*

46 Denis Dodart, "Mémoire pour Servir à l'Histoire des Plantes," in *Mémoires de l'Académie Royale des Sciences depuis 1666 jusqu'à 1699,* Tome IV (1731), 121–333; Claude Perrault, *Mémoires pour Servir a l'Histoire Naturelle des Animaux* (Paris, 1671).

47 See also Holmes, *op. cit.* note 14, 177–178.

48 While translations of articles by Robert Boyle and other English authors did appear in *Journal des Sçavans,* it is worth noting that our complete random sample of French whole articles and short passages contains only two authors not members of the Académie Royale (Francine and Bernoulli), neither of them English.

49 Ben-David, *op. cit.* note 9, 108.

50 This early professionalization, as well as its financial underpinnings, has been masterfully documented by historian Alice Stroup (*op. cit.* note 4).

51 Not surprising but sad to say, this openness did not extend to women authors. Atkinson (*op. cit.* note 13, 102) reports that the first article with a woman listed as author did not appear in *Philosophical Transactions* until 1760: "Effects of a Thunderstorm at Rickmansworth, Herts," by Ann Whitfield. The astronomer Caroline Herschel reported her sightings of comets in several short *Philosophical Transactions* articles published in the 1790s.

Transactions delivered on the promise of its title page: "Giving some accompt of the present undertakings, studies, and labours of *the ingenious in many considerable parts of the world*" (italics added). And the articles therein favoured the reporting of observationally derived facts *sans* theoretical explanations.

It is also worth bearing in mind that, in the French case, the relationship between professionalization and scientific explanation is not straightforward. It is not that we have, as a direct consequence of this narrowing, a Cartesianism that insists science deliver a material cause, or the Galilean view that the book of nature is necessarily "written in the language of mathematics." Rather, this narrowing—coupled by the talents of such as Huygens or Varignon—opens one set of opportunities for French science while closing off another. While the English are open to "natural science as a model for valid thinking about political, economic, and technological matters,"[52] the French pursue a narrower goal with singular intensity. Between this French pursuit of narrower goals and instrumental precision, the path is direct, a concern for such precision being the only means of linking theory firmly to observations and experimental results. During the 18th century, we suspect, these interesting differences will become less pronounced as French science diversifies with the proliferation of scientific journals and societies,[53] while the English scientific article comes more closely to resemble the French example under the powerful sway of Isaac Newton and his disciples.[54]

Appendix: Questionnaire for Argumentative Practices

Questions asked of randomly selected whole articles, 50 English and 50 French, except for "hedging" and "quantitative" statements, where questions asked of randomly selected ten-line passages, 100 English and 100 French.

Verbal Argument

1. Is the article's content (a) experimental? (b) theoretical? (c) methodological? (d) observational? (e) observational/theoretical? (f) experimental/

52 Ben-David, *op. cit.* note 9, 108.

53 McClellan, *op. cit.* note 4, 67–68, 89–99.

54 For good discussions of the argumentative practices in the 18th-century *Philosophical Transactions,* see this work cited in note 13: Bazerman, 59–79; Valle; and Atkinson, 75–109.

theoretical? (g) experimental/methodological? (h) medical/surgical? (i) mathematical?

2. Is the article's main claim an (a) observation? (b) experimental results? (c) mechanical explanation? (d) mathematical rule? (e) mathematical explanation? (f) making or improving equipment?
3. Is there any mention of witnesses to an observation or experiment?
4. How many hedging expressions are there per 100 words? Examples include: "il y a des conjonctions qui **paroissent** précises" (1692-MAS080); "mon sentiment, **s'il est vrai**, perfectionne ce qu'il y a **de général** dans son Systême" (1699-MAS051); "that those veins do any where grow less and smaller, though **probably it may be so**" (1672-PT063).
5. How many quantifying expressions and equations are there per 100 words? Examples include: "A **2 heures 48'**, la Lune étoit éclipsée de **9 doigts 58''**" (1692-MAS041); "Soient enfin **AE = x, EB = y, & EF = z**" (1699-MAS067); "It appears that in the **Five Years** mentioned, viz. from **87 to 91** inclusive, there were *born* **6193 Persons**, and *buried* **5869**" (1693-PT041).

Visual Argument

1. How many graphics in articles (tables and figures)?
2. What is percent of space of graphics (tables and figures)?
3. If there is a graphic, is it of (a) equipment? (b) experimental arrangements? (c) natural object? (d) is it a list? (e) a table? (f) a mathematical relation?
4. If there is a graphic, does it support (a) an observation? (b) an experimental result? (c) a mechanical explanation? (d) a mathematical relation? (e) a mathematical explanation? (f) improving equipment?

PART 2

Case Studies

CHAPTER 5

Invitation and Engagement

Ideology and Wilkins's Philosophical Language

Robert E. Stillman

As early as October of 1662, the Royal Society was urging "Dr. Wilkins…to prosecute his design of an *universal language*."[1] John Wilkins was among the Society's most valued communicators: together with Henry Oldenberg, he was appointed one of its two secretaries, and was later assigned to its Correspondence Committee. Asked to supervise the Committee for Improving the English Tongue, Wilkins reviewed its work with an eye to remedying "the philosophy of the language."[2] Other fellows—including John Beale, Seth Ward, Robert Hooke, and William Petty—had a strong interest in the creation of a universal language, but Wilkins was an especially apt choice for this Society project, for by 1662 his *Essay Towards a Real Character and a Philosophical Language* was almost certainly work-in-progress. Confronted by the challenge of having to justify its newly institutionalized claims upon the public interest, the Royal Society identified the construction of a philosophical language as a primary goal and chose Bishop Wilkins as the appropriate man to pursue that goal's realization.[3]

I begin by situating Wilkins's *Philosophical Language* inside the institutional context of the early Royal Society. I do so as a necessary supplement to traditional readings of the essay that strive to interpret his project largely (sometimes exclusively) against the background of the universal language movement. The story of that movement has been well told by Hans Aarsleff, Vivian Salmon, James Knowlson, and M.M. Slaughter.[4] Follow its course as it develops in the

1 Thomas Birch, *The History of the Royal Society of London* (New York. Johnson Reprint, 1968), vol. 1, 119 (October 29, 1662).

2 *Ibid.*, vol. 2, 7.

3 On the many projects designed to acquire greater financial credit and enhanced public credibility for the Society, see Michael Hunter, *Establishing the New Science. The Experience of the Early Royal Society* (Woodbridge, Suffolk: Boydell Press, 1989). For the importance of communications to the early Society, see Peter Dear, "*Totius in verba*. Rhetoric and Authority in the Early Royal Society," *Isis* 76 (1985), 145–161.

4 For an account of that movement in a larger diachronic analysis, see the influential study by Hans Aarsleff, *From Locke to Saussure. Essays on the Study of Language and Intellectual History* (Minneapolis: University of Minnesota Press, 1982). See also Vivian Salmon, *The Works of Francis Lodwick. A Study of His Writings in the Intellectual Context of the Seventeenth*

 | DOI 10.1163/9789004283701_007

wake of Bacon's call for a philosophical grammar, from the publications of real characters to the production of universal languages, and much that is crucial to understanding Wilkins's essay emerges: from his particular debts to George Dalgarno in shaping his real character, to his more general borrowings from Mersenne in shaping his attitudes about language.[5] Taken as a whole, however, the seventeenth-century universal language movement is too diversified, too factionalized among what Humphrey Tonkin has called philosophical sheep and mystical goats, to explain the specific motivations inspiring Wilkins's *Philosophical Language*.[6] Wilkins articulates goals broadly familiar from that movement—the enhancement of commerce, the clearing of religious differences, the development of a "scientifically" exact vocabulary; but none of those goals is sufficiently precise to explain why the essay achieves its particular organization and character as the distinctive product of a particular historical

Century (London: Longman, 1972); idem, *The Study of Language in Seventeenth-Century England* (Amsterdam: Benjamins, 1979); James Knowlson, *Universal Language Schemes in England and France, 1600–1800* (Toronto: University of Toronto Press, 1975); and M.M. Slaughter, *Universal Languages and Scientific Taxonomy in the Seventeenth Century* (Cambridge: Cambridge University Press, 1982).

Wilkins's essay has also been of interest to contemporary linguists. For a review of some recent claims by contemporary linguists for the continued value of the essay, see Joseph L. Subbiondo, "John Wilkins's Theory of Meaning and the Development of a Semantic Model," *Cahiers linguistiques d'Ottawa* 5 (1977), 41–62. See also Fredric T. Dolezal, "John Wilkins and the Development of a Structural Semantics," in *Papers in the History of Linguistics*, eds. Hans Aarsleff, Louis F. Kelly, and Hans Josef Niederehe (Amsterdam: John Benjamins, 1987), 271–281, and Dolezal; *Forgotten but Important Lexicographers. John Wilkins and William Lloyd; A Modern Approach to Lexicography before Johnson* (Tubingen: Max Niemeyer Verlag, 1985).

More recently, some critical attention has been given to the essay's historical situation inside Restoration culture at large. For suggestive commentary about the relationship between Wilkins's essay and "the significance of writing in the evolution of bourgeois society," see Tony Davies, "The Ark in Flames. Science, Language, and Education in Seventeenth-Century England," in *The Literal and the Figural. Problems of Language in the History of Science and Philosophy, 1630–1800*, eds. Andrew E. Benjamin, Geoffrey N. Cantor, and John R. R. Christie (Manchester: Manchester University Press, 1987), 83–102; and for an analysis of Wilkins's linguistic proposals as a "scientific" effort to restore Adam's fusion of signifier and referent, see Joel Reed, "Restoration and Repression. The Language Projects of the Royal Society," *Studies in Eighteenth-Century Culture* 19 (1989), 399–412.

5 See David F. Cram, "George Dalgarno on 'Ars Signorum' and Wilkins's 'Essay,'" in *Progress in Linguistic Historiography*, ed. Konrad Koerner, Studies in the History of Linguistics, series 3, vol. 20 (Amsterdam: John Benjamins, 1980), 113–122.

6 Humphrey Tonkin, "Real Character and Lunatic Language. Language Planning in the Seventeenth and Eighteenth Centuries," *Language Problems and Language Planning* 3, 2 (1979), 78.

moment. I will argue that only in terms of the historically specific circumstances of Restoration culture—in light of the political concerns of the Royal Society and its public campaign to invite and engage support for the new philosophy—is it possible to explain why Wilkins's language masters the stuff of history into the particular shape it achieves.[7]

Wilkins's longtime immersion in the universal language movement was a consequence of his deepest, most abiding interests. The universal language was "his Darling," as John Aubrey reported, and "nothing troubled him so much when he dyed, as that he had not completed it."[8] As a natural philosopher working to rescue real truths from "worded disputations," Wilkins heralded the language as a vehicle to "promote and facilitate the knowledge of nature."[9] As a bishop, struggling to unmask the "wild errors" of wild words, he advertised its advantages of "clearing of some of our Modern differences in Religion" (Epistle Dedicatory, n.p.). As a citizen of London, he promoted its benefit "of facilitating mutual *Commerce* amongst the several Nations of the World" (*ibid.*). And as a mighty rising man among the Restoration's elites, keenly aware of "all the unhappy consequences" of "the Curse of the Confusion," especially those "many impostures and cheats that are put upon men, under the disguise of affected insignificant Phrases," he celebrated the power of philosophical commerce ("the genuine and natural importance of Words") to undo some of Babel's most nefarious consequences in the wider political domain (*ibid.*). Finally, and most importantly, as a fellow of the Royal Society dedicating his work to the Society, Wilkins emphasized the "abundant Interest" his language

7 For an early effort to define Wilkins's aims in Comenian terms, see Benjamin Demott, "Comenius and the Real Character in England," *PMLA* 70 (1955), 1068–1081. In response, see Vivian Salmon, "Language-Planning in Seventeenth-Century England. Its Contexts and Aims," in *In Memory of J.R. Firth*, eds. C.E. Bazell, J.C. Catford, M. A.K. Halliday, and R.H. Robins (London: Longmans, 1966), 370–397; while Salmon is correct to be suspicious of identifying Wilkins's aims with Comenius's, her claim that he is "primarily a scientific linguist" begs important questions about precisely what it means in the England of the 1660s to be a *scientific* linguist (p. 377). I would argue that such claims (to the status of scientist or natural philosopher) always and necessarily entail—given the Society's practical need to justify its share in the public interest—an ideological content.

8 John Aubrey, *Brief Lives*, ed. Oliver Lawson Dick (Ann Arbor: University of Michigan Press, 1957), 320.

9 John Wilkins, *An Essay Towards a Real Character and a Philosophical Language* (London: Samuel Gellibrand and John Martyn, 1668), Epistle Dedicatory, n.p. (All further citations to Wilkins are from this edition and are documented in the text parenthetically by page number.)

would pay the institution itself, with its "universal benefits" to natural philosophy, religion, commerce, and politics (*ibid.*).

The Society, in turn, had its own powerful motives for promoting Wilkins's language. The *power* of those motives is suggested both by its official encouragement of the project and by its provision of the official Society imprimatur to its publication. Wilkins's *Philosophical Language* is a shrewdly crafted exercise in institutional propaganda, a Society production in praise of Society values. At its center is the celebration of natural philosophy as a vehicle for the accumulation and exchange of knowledge. As Thomas Sprat heralds the Society as "the general *Banck*, and Free-port of the World," an idealized entrepôt of philosophical commerce, so Wilkins praises its work in facilitating a more perfect exchange of knowledge to enable a more useful exchange of socially valuable goods of other sorts, from the tangible goods of commerce to the less tangible goods of moral, religious, and political concord.[10] Earlier in the 1660s, other Society texts—such as John Evelyn's *Sylva*, John Graunt's *Observations*, and Hooke's *Micrographia*—had advertised themselves as contributions to that commerce, and as the tentative "Towards" in Wilkins's title indicates, his text is likewise to be measured as a *contribution* to a larger collective enterprise.[11]

It is a contribution, however, more ambitious than the earlier publications of Evelyn, Graunt, and Hooke. If the Society was to become a center of calculation, a storehouse of new philosophical learning, then Wilkins's *Philosophical Language* struggles to occupy the center inside that center—not as an accumulation of knowledge in a specific domain, but as a vehicle for mobilizing and exchanging knowledge across a universe of domains. By furnishing a set of universal symbols coincident with the universals of philosophical knowledge, the language strives to enable, in a more perfect commerce of symbols, a more perfect commerce of things. Moreover, with the familiar seventeenth-century extension of idealized symbolic harmony to social harmony, the agency of the well-made language in facilitating the well-made state, at the core of Wilkins's *Philosophical Language* is an equally familiar desire to master history by mastering signs.[12] His language, in short, is comprehensible not merely as one

10 Thomas Sprat, *History of the Royal Society*, eds. Jackson I. Cope and Harold Whitmore Jones (Saint Louis: Washington University Studies, 1958), 64.

11 For a documentary account of the Society's early publications, see Charles A. Rivington, "Early Printers to the Royal Society, 1663–1708," *Notes and Records of the Royal Society of London* 39 (1984), 1–27.

12 See, for example, Stephen Shapin and Simon Schaffer's discussion of the Society as "a model of the ideal polity," in *Leviathan and the Air-Pump. Hobbes, Boyle, and the Experimental Life* (Princeton: Princeton University Press, 1985), 341.

more expression of the Society's official program for celebrating the social importance of its own work, but also as the very means by which that expression could be realized inside the culture at large. Let me explain.

Wilkins's essay is divided into four main sections. The first part is a Prolegomena, which establishes the need for a philosophical language. The second is an account of Universal Philosophy, a provision of tables "to reduce all things and notions unto such a frame, as may express their natural order, dependence, and relations" (p. 1). In part three, Wilkins constructs a "natural" grammar for forming the "more simple notions" of the Philosophical Tables "into complex Propositions and Discourses" (p. 297). In part four, he supplies his real character, as a set of written signs referring to the things of the Tables and the relations of the grammar, and directions for making that character "effable in a distinct language" (p. 414). The essay is organized as a grand symbolic entrepôt: in the Philosophical Tables, Wilkins accumulates into a massive repository all things and notions, methodically framed in a universe of predictable relations; in the grammar, he provides the rules for mobilizing those things and notions in a logically governed discourse; and in his descriptions of the real character and the effable language, he affords the technical means for making that mobilization of philosophically defined things and notions possible. From the outset, as he demonstrates the need for a philosophical language, it is apparent how urgently his new philosophical commerce is fueled by social concerns.

Wilkins's Prolegomena forms a critique of "the various Changes and Corruptions, to which all vulgar Languages are obnoxious," a critique whose ferocity and comprehensiveness are unparalleled in seventeenth-century England since Bacon (Contents, n.p.). In the footsteps of his master, Wilkins turns the full weight of centuries of humanist philological learning against the adequacy of language itself.[13] As his chief wedge for reopening a Baconian gap between words and things, he introduces a single, all-purpose principle: "every change is a gradual corruption" (p. 8). Measured by that principle, all the historical sophistication of humanist learning, in its demonstration that language is the child of time, amounts to nothing short of a massive indictment of the child's inadequacies. Linguistic change is corruption; corruption is the work of time; and time's most nefarious manifestation is "general Custom," the mutability of history itself (p. 18).

Custom's nefarious manifestations are detailed with painstaking completeness. Alphabets change: "the *Letters* of any Tongue do not alwaies remain the

13 For a useful study of Bacon's relationship to the humanist tradition, see Charles Whitney, *Francis Bacon and Modernity* (New Haven/London: Yale University Press, 1986).

same, but are subject to...fate and mutability" (p. 12). As a consequence, "their *Powers*...are very uncertain, not alwaies fixed and determined to the same signification," spawning by alphabetical mutations what Wilkins calls in one of the century's favorite tropes "monstrum literae" (pp. 15–16). He details, too, the defects of words in their "Accidents and Constructions," noting their equivocal redundancy and complaining especially, in the familiar fashion of the new sciences, about "the ambiguity of words by reason of *Metaphor*" (p. 17). Metaphors are dangerous because they produce "false appearances" and "are very changeable, every generation producing new ones" (p. 18). Witness "the late times," Wilkins writes, in one Restoration version of the civil wars as England's recapitulation of Babel, when "these Canting forms of speech...hath almost eaten out solid Knowledge in all professions" (p. 18). Canting words are a cannibalization of knowledge. Pronunciation changes, too, as an illustration that Custom has "rivetted...incongruity and imperfection in all Languages" (p. 18). Once more, orthography alters from generation to generation, as Wilkins laments that Custom is "so invincible...that still we retain the same errors and incongruities in writing which our Forefathers taught us" (p. 19). Mutations in the parts bring corruption to the whole—both for the most venerable of languages, Hebrew, and for English, whose accelerating pace of change in its "addition of new words" signals accelerating decline. Language is no neutral medium for Wilkins: it is a dark romance of false appearances, peopled by monsters, cannibals, and canters threatening, with fearsome cultural consequences, to deform knowledge itself.

Against the complex of linguistic and political problems created by "general Custom," Wilkins's solution, as first articulated, seems disarmingly simple. Alphabets mutate, pronunciation alters, equivocations abound as custom reigns, but philosophy can undo those changes, since beneath the flux of words "men do generally agree in the same Principle of Reason, so do *they* likewise agree in the same *Internal Notion* or *Apprehension of things*" (p. 20). Given the existence of universal notions and their reliability, Wilkins reasons that by assigning *Marks* to "our Conceptions" it is possible to "signifie *things*, and not *words*" (p. 21). Perceived in these terms, Wilkins's "chief difficulty and Labour" appear technical in kind—the discovery of "adequate" signs and the avoidance of redundancy—but when he considers how such signs are to be made "*regular* as to their Place and Order," far greater complications trouble what seemed originally a simple solution (p. 20).

To give signs their Place and Order, as Wilkins well knows, presupposes a comprehensive knowledge, philosophically verifiable, of the Order of the universe: "the *Theory* it self, upon which such a design were to be founded, should be exactly *suted to the nature of things*"—otherwise, he adds ominously, there "must needs [be] much perplexity" (p. 21). In the face of perplexity and the fearsome threats of custom, what seems called for here is an epistemological argument securing the grounds of that order; yet no such argument appears here or elsewhere in the essay. The best that Wilkins can achieve is a pragmatic solution to an epistemological problem. Diffident about entering into the argumentative fray, in line with a lifelong modesty that rebels against premature conclusions, he recognizes that in the absence of an absolutely verifiable theory of knowledge, a philosophical language can spring into being only as a consequence of two historical possibilities. First, it might come about because "some one Language and Character" will be enjoined "to be universally learnt and practised, (which is not to be expected, till some person attain to the *Universal Monarchy*...)" (like so many Restoration thinkers, Wilkins is haunted by the specter of *Leviathan*: Universal Monarchy is as repellent and impractical an idea as the epistemological absolutism that might lead to its creation). Or second, a universal language might come into being because "some such way as, by its facility and usefulness" will be proposed "(without the imposition of Authority) [that] might *invite* and ingage men to the learning of it" (p. 20).

An authoritative linguistics enjoins. A more open and engaging linguistics invites. The first alternative, threatening dogmatism and absolutism, is Hobbes's, as the philosophical consequence of an authoritarian politics aimed against the monsters of the civil wars.[14] This is an alternative that Wilkins both explicitly considers and quickly rejects. The second alternative is Wilkins's own, a philosophical linguistics founded upon "common assent" about universal principles of reason, leveled against the nefarious consequences of general custom. Accommodated to a model of political consensus, as I will show, it is a linguistics scrupulously designed to harmonize with the Restoration gentry's ideal of a free commerce of negotiable rights, contracts, and obligations. Wilkins's solution to his epistemological problem, then, is ideological as well as pragmatic. It can be described as ideological because it strives to impart natural justification to a historical intention, to lend the appearance of eternality and universality to what are demonstrably the contingent values of a Restoration elite.

Wilkins's real work of inviting and engaging begins in the essay's second part, with the provision of the Philosophical Tables. Enter the linear space of these tables, conveniently reduced to two giant-sized folio pages, and there appears an entire universe of simple things and notions (pp. 442–443). (They

---

14 I expand upon this characterization of Hobbes and his relationship to Wilkins in a forthcoming essay: in Robert E. Stillman, "Hobbes's *Leviathan*. Monsters, Metaphor, and Magic," *ELH* (in press).

are "simple" things because it is upon the basis of their more complex permutations and combinations that the larger universe is constructed.) As Slaughter has ably demonstrated, Wilkins employs an Aristotelian taxonomic scheme for ordering the universal domain of "things"—from objects to notions, nature to culture—in a hierarchy descending from the most general categories to the most particular properties.[15] The Philosophical Tables detail relations between individual things and groups, between subordinate groups and major groups, in such a way as to configure (ideally) a complete network enumerating the sum total of relations among the universe of things and notions. Knowledge about any one thing leads directly, through the predictive agency of the Tables, to a knowledge of its relationship to all other things. Once mark those things by legible characters and effable signs, and determine logical pathways for mobilizing those things into discourse, and the universal language enables a commerce of universal knowledge.

As an organizing hierarchy, Wilkins creates forty Genuses, "or more common heads of things belonging to this design" (p. 22). (Forty is an arbitrary, albeit biblically evocative, number: while Wilkins's chief contemporary model for the language is the entrepôt, his chief scriptural model is Noah's ark, with its similarly salvific, economically designed, and comprehensive accumulation of things. To Pepys's delight, Wilkins devotes a full digression to the ark.) Each Genus is subdivided "by its peculiar *Differences*," generally limited for "convenience" to six; and each of these Differences is divided into several Species, "according to such an order and dependance amongst them, as may contribute to the *defining* of them, and determining their primary significations" (p. 22). All things belong to a particular Genus because they share a "common Essence," and their organization into Differences and Species provides additional means for specifying their expression of that essence. Each "thing" in turn is paired with a contrasting or complementary thing. Two by two, the creatures assemble inside a linguistic ark to weather, amidst its forty Genuses, the troubled seas of Restoration crises.[16]

Slaughter is correct to characterize this taxonomic scheme as Aristotelian, and to emphasize its usefulness in facilitating a comprehensive ordering of things—but clearly, too, part of the scheme's appeal to Wilkins lies in its power to invite and engage. However novel or even bizarre his characters and effable

15 Slaughter, *Universal Languages* (above, n. 4), 167–177.

16 On the Restoration's crisis of legitimation, and the urgency of the culture-wide search for authority from the domain of language-planning to that of politics, see Gerald M. MacLean, *Time's Witness. Historical Representation in English Poetry, 1603–1660* (Madison: University of Wisconsin Press, 1990), 259.

signs appear, the universe he constructs is reassuringly familiar. No educated Restoration reader—and it is chiefly the educated to whom Wilkins appeals—would have been puzzled, for instance, to find that a stream is a Species of Water, and that Water, in turn, as one of six Differences, belongs to that still more comprehensive Genus, the World. Nor would that reader have been surprised to find, according to the same scheme, that a divine is a kind (or Species) of Profession, and that Professions, in turn, as one of six Differences, belong to that more comprehensive Genus, Relations Civil.

Wilkins's scheme, then, engages by virtue of its confirmation of the traditionally approved wisdom of the university-educated—still grounded in Aristotelian-Scholastic learning—just as it invites by virtue of its comfortably conservative world picture. As in Browning's poem, God's in his heaven, and all's right with the world. Rub the gloss from this philosophically designed taxonomy, and what shines forth are the none-too-burnished links of that hoariest of all Western metaphysical models, the Great Chain of Being: among the forty Genuses, Wilkins moves up this chain from nature to society, from the constituent parts of the world (from the inanimate to the animate) to the constituent parts of culture (from family relations to civil, judicial, military, naval, and ecclesiastical relations). Every "thing" and notion finds its place in a universal order, as the expression of a "common Essence," and the same essentialism that stipulates the rightness of placing streams among kinds of water also stipulates the rightness of placing divines among kinds of professions. By means of this essentialist metaphysics, Wilkins creates across the domains of nature and culture a fully commensurable universe of things, as if the ineluctable principles establishing natural hierarchies are the same principles establishing social hierarchies. Once more, the taxonomic space of the Tables is not to be imagined as neutral or merely convenient; its hierarchies are everywhere determinative of both meaning and value, in its gradations from the inanimate to the animate, from the natural to the human, and often even in the gradations imposed among individual Species and Differences. The divine's priority to the player, among the Species of Professions, acquires the appearance of an essential—not a taxonomically convenient—rightness, parallel to the right subordination of streams to the ocean.

Wilkins's Tables mark meaning and value sometimes, too, not just for the things they include, but also for those things that they specifically exclude. The fact that the Tables are designed to exclude certain vast areas of experience comes as something of a surprise to the reader, especially as that exclusion is advertised only toward the end of part two. No characters are to be assigned for such things "as are appropriated to particular *Places* or *Times*" (p. 295). Titles of honor and office, degrees in professions, legal and heraldic terms, games and plays, fabrics, drinks, meats, tunes, and tools, all are excluded from the real character and the philosophical language. Last on the list of excluded items are "the names of divers sects, whether Philosophical, Political, or Religious" (p. 296). Wilkins is not especially apologetic about these acts of exclusion, and it would be wrong to interpret them as a sign of tentativeness in an essay only gesturing *toward* a philosophical language.[17] Instead, they are to be judged as a fully deliberate gesture consistent with Wilkins's stated intention to rescue language from the hitherto invincible assaults of general custom. What is merely mutable, customary, and historical does not belong in a philosophical language, contrived according to "a Theory of things as is proposed for Universal use" (p. 296). This does not mean that the user of Wilkins's language cannot refer, for instance, to a particular philosophical, political, or religious sect (a periphrastic expression of its name in Latin or the vernacular serves well enough). What it does mean, by the very exclusion of that name from the character, is that that particular sect will immediately be marked as being *merely* historical or *merely* customary, relative to the universal things and notions of the philosophical language.

As Slaughter has indicated, then, Wilkins's commitment to Aristotelian essentialist taxonomy secures the organizational framework of his *Philosophical Language*. An accumulation of things, methodically organized in predictable relations, comprises the first step in creating a center of calculation. In order to appreciate more fully the essay's power to invite and engage, however, it is necessary to perceive how Wilkins persuades his readers to participate in a distinctively, if peculiarly, idealized form of Restoration commerce. He affords not just a comfortably conservative worldview, but also—as a shrewd Society spokesman—a worldview specifically tailored to the tastes of his culture's sovereign elites. In the distinctive fashion of the Restoration's public discourse, his essay circles persistently around an economic vocabulary of credit, credibility, and interest.[18] What is inviting about his recurrent employment of that vocabulary is its translation into a philosophical framework that rescues interested pursuits from the perilous ambiguities of ordinary language and life.

17 For a different emphasis, see Sidonie Clauss, "John Wilkins's Essay toward a Real Character. Its Place in the Seventeenth-Century Episteme," *Journal of the History of Ideas* 43, 4 (1982), 531–554.

18 On the pervasiveness of the Restoration's vocabulary of political economics, see Albert O. Hirshman, *The Passions and the Interests. Political Arguments for Capitalism before Its Triumph* (Princeton: Princeton University Press, 1977); and J.A.W. Gunn's more dated study, *Politics and the Public Interest in the Seventeenth Century* (London: Routledge and Kegan Paul, 1969).

Consider the place that Wilkins accords his language-user as a potential actor amidst the universe of things. Defined as an intellectual being, man is possessed of speculative understanding similar to the spiritual action of God, lower in status but essentially similar in nature. As God's spiritual action extends in its Species from creation and providence to revelation and redemption, man's speculative understanding proceeds from thinking and inquisition to reasoning and esteeming ("accounting, prizing, valuing, ...credit, prefer" [p. 227]). The goal of God's spiritual action is the work of salvation; the implied goal of man's speculative activities, in thinking, assenting, and believing, is the work of esteeming. Giving credit is not the socially learned behavior of a Restoration gentleman, Wilkins's scheme indicates; rather, it is a universal and innate intellectual activity allied to the actions of the divine.

So enabled, the image of man's speculative understanding as defined by the Philosophical Tables is optimistically appealing. Paired two by two with sometimes negative complements (certainty with mere opinion, reasoning with frivolous conjecturing), the Species of speculative understanding show Wilkins's real awareness of the defects to which intellectual activity is subject, but there is nothing inherent in human understanding to prevent the mind from arriving at appropriate acts of esteem. Bacon's idols are inconvenient and forgotten shibboleths. Esteeming is an essential function of the speculative understanding allied across the Genuses to man's natural powers of reasoning, "whereby we are inabled to apprehend and compare the general natures of things as to Truth and Falsehood, Good and Evil, and to demean our selves accordingly towards them" (p. 195). One step up the ladder, science is defined among the acquired intellectual virtues, as a more precise rational tool "furnishing the mind with due Notions and conceptions concerning the Nature of things, their Causes, Differences, Relations, and Dependencies" (p. 205).

Reason is an action of the speculative understanding, linked to esteem; a natural power that illuminates (for the *general* nature of things) truth and falsehood, good and evil; and an intellectual skill enhanced by science to render esteem more certain. Use the philosophical character for maiz, and one knows immediately that maiz is a kind of "frumentaccious" herb, classified according to leaf; use the character for humility, and one knows immediately that humility is a kind of manner adopted toward superiors. The very tables of Wilkins's *Philosophical Language*, with their comprehensive display of the order of things, stand as palpable evidence of acquired intellectual virtues at work, and as a palpable means for the user, freed from the distortions of inadequate languages and inadequate metaphysics, to acquire reason in appropriately crediting the truth and falsehood, the good and evil of things themselves.

While esteeming (giving credit) defines one essential intellectual activity, reputation (gaining credit) defines one major goal of social activity. Under the Genus of Habit, Wilkins lists "REPUTATION, Credit, Countenance, Applause," as one Species of those instruments of virtue "whereby the natural Faculties are perfected" (pp. 200–201). Always, he is concerned to forge connections between principles and duties, between natural faculties and social virtues, and reputation stands as one main means to make that connection. Although it is not itself a virtue, "the esteem we have amongst good men" is defined as "requisite to the due exercise of the Acts of many Vertues, and one kind of Reward belonging to it" (p. 201). (Note that esteem is gained only from *good* men—and only the esteem of the good, the assent of the best, counts.) Among the other instruments of virtue, Wilkins accords Reputation a central place, mediating between liberty, riches, and pleasure (archetypal gentry values) and dignity and power (archetypal gentry goals). Power is aligned with interest as "our Ability to protect our selves and others from injury," and it is acquired by liberty, riches, reputation, and dignity (p. 201). Properly understood and appropriately defined, to pursue one's interest is to enable a social commerce of virtue. In a transparent elevation of the contingent values of his Restoration culture to the status of transcendentals, Wilkins enshrines the pursuit of credit as a universal social goal in parallel fashion to his designation of credit as a universal speculative activity.

Reputation stands as one chief value in philosophical tables representing society as a marketplace for the exchange of real and symbolic capital. To the Restoration's familiar celebration of "commerce doux," Wilkins provides more than a local habitation and a name: his language generates, more comprehensively, a philosophical lexicon that imparts natural legitimation to the civil exchange of goods and knowledge—a lexicon both inviting, in its emphatic provisions for "free and generous men" to pursue their "RIGHTS, Title, Interest, Estate," and engaging, in its equally emphatic provisions detailing the "OBLIGATIONS" of those men (p. 267). At the foundation of this representation of society is the familiar and familial logic of the patriarchy. Under the Genus of Oeconomical Relations, Wilkins describes the family as "the first and most natural kind of association of men" (p. 249). The natural quality of this social construct is crucial because it underwrites the legitimacy of those two greater activities that organize the comprehensive network of civil relations: the pursuit of possessions and provisions, as things necessary to the "well-being of a Family," and the maintenance of degree, "Relations of

SUPERIORITY...and INFERIORITY, *Underlings*, do originally belong to Families, in which was the first kind of Government" (p. 254, 250).[19]

A hierarchy determines the relative superiority inside the family among those who provide for souls (godfathers), minds (teachers), and estates (guardians)—just as education, which concerns "the due Government of persons," proceeds from commanding and persuading to commending and praising in order to win assent for that hierarchy's determinations (p. 252). The organization of society, like the organization of language, is a pragmatic business, which works toward common assent by inviting and engaging. Move from economical relations to civil relations, and a parallel hierarchy organizes political degrees from magistrates, kings, and lords to the vast populace beneath.

In a telling example of Wilkins's effort to enable a social commerce that operates by inviting and engaging, his magistrate is defined as "such to whom doth belong the power and care of directing others in their duty, and rewarding or punishing them accordingly" (p. 264). Represented as the guardian of duty, who rewards and punishes, invites and engages, credits and contemns, the magistrate safeguards a society organized, not to maintain his power, but to regulate the commerce of free and generous men. The essential "*Things* or businesses" of civil relations concern Rights, Contracts, and Obligations: while rights devolve from nature, custom, law, property, and prerogative ("Royalty"), contracts provide "for the Alienation of our Rights," and obligations "for the Confirmation of Contracts" (p. 263). So conceived, the center of civil society belongs to EXCHANGE, wherein "The mutual negotiating betwixt men in their ordinary converse with one another" secures "the most general occasion for such Intercourse, namely, the parting of one thing for the getting of another" (p. 268). *Exchange, commerce, converse*, and *intercourse*: everywhere the vocabulary of economics overlaps with the vocabulary of communications, fueling a cycle, at once material and symbolic, of parting with one thing for the getting of another. As a civil extension of the family's natural pursuit of provisions, the business of accumulation and exchange defines, not the emerging capitalist activity of a Restoration culture (or so his argument would have it), but the essential activity of universal civil relations.[20]

19 Still the best study of patriarchalism, and the extension of notions about natural law and the family to political authority, is Gordon J. Schochet's *Patriarchalism in Political Thought. The Authoritarian Family and Political Speculation and Attitudes Especially in Seventeenth-Century England* (New York: Basic Books, 1975).

20 On the intersection between the financial and the linguistic domains in seventeenth-century representations of commerce, see Jean Christophe-Agnew, *Worlds Apart. The*

Extending the magistrate's authority into institutional hierarchies as a safeguard for that activity, legal, military, and naval relations add coercive power to the state's persuasions to assent. Such institutions are the universal complements of social commerce. Adapted to civil relations, Wilkins's language is tailor-made for the emerging fiscal-military state,[21] and nothing displays the hand of the tailor more clearly than the ready "mesh" of its fiscal-military vocabulary with the signs of religious devotion. In a latitudinarian display of ecclesiastical relations, Wilkins incorporates a diverse vocabulary of persons, conditions, and forms of worship inculcating a "habit of reverence towards the Divine nature" (p. 284). As man's speculative understanding terminates in acts of esteem, his civil relations end in "That inward and outward reverence whereby we acknowledge the Esteem due to the Superiority and Excellency" of God (p. 286). In a self-authenticating cycle of relations, from essential operation of mind to essential activity of society to essential function of worship, credit, as Wilkins once sermonized to King Charles, "Rules the World, and Stamps upon Things the Rates at which they are to pass."[22]

Across the comprehensive expanse of the Philosophical Tables, as Wilkins moves from nature to culture, from principles to duties, there are revealing shifts in standards of credit. Considered in terms of intellectual activities, Wilkins's speaker, as I have indicated, is remarkably well empowered: the speculative work of esteeming aids in saving moral certainty from opinion, reality from appearance, as the power of natural reason aids in distinguishing, among the general nature of things, truth from falsehood, good from evil; further enhanced by "science," reason can penetrate the nature of things to determine their probable causes, differences, relations, and dependencies. Considered in terms of social activities, however, Wilkins's speaker is considerably less powerful. In civil relations, reason as a natural power gives way to "right reason," that faculty governing "our Actions towards others, in such cases as the Lawgiver (could he have foreseen) would have provided for" (p. 207). In society, the individual gives up "his own strict right…to take the most amicable way in the accommodating of Differences" (p. 207). Reason about the nature of things is replaced, in the cause of amicable social relations, by obedience to law or by reason that accords with the consent of the best, with the judgment of the

---

Market and the Theater in Anglo-American Thought, 1550–1750 (Cambridge: Cambridge University Press, 1986).

21 See John Brewer, *The Sinews of Power. War, Money, and the English State, 1688–1783* (London: Unwin Hyman, 1989).

22 John Wilkins, *A Sermon Preached before the King upon the Seventh of March 1668/9* (London: T. Newcomb for Sa. Gellibrand, 1669), p. 43.

sovereign or of the sovereign's representatives. Virtues accord not with the dictates of conscience, but with "Duty" or the "Esteem" of enhancing ESTATES and DIGNITIES (pp. 208–209). Passions are judged good or evil, not by the natural power of reason, but "with respect to the particular interest of *Repute*"—always the repute to be acquired among good men (p. 232). Social commerce is so regulated to subordinate individual judgments to standards of community assent: the best are empowered to credit what is true. And the best are so empowered through the metaphysical guarantees of the Philosophical Tables, not because they are the powerful (or so Wilkins would have it), but because such acts of universal assent accord with the universal nature of things. The *Philosophical Language* achieves, then, a distinctive marriage, at once pragmatic and ideological, between truth as a social construction and truth as a natural given.[23]

In the absence of an absolutely verifiable theory of knowledge, Wilkins appeals for the adoption of his language by virtue of "its facility and usefulness" (p. 20). His essay stands as an invitation to a contract, an appeal for common assent not just about a new order of real characters and effable marks, but also about a philosophical framework fixing the semantic boundaries within which signs negotiate meaning and value. Much of his essay's power to invite and engage, as a Society project in pursuit of Society credit, derives from its provision for the sort of commerce with which any Restoration gentleman would have been pleased. It lends access to a hierarchically organized universe of things, conveniently displayed as a kind of lexicon for virtuosos to facilitate both knowledge and communication. Accommodated to a society whose civil relations are governed by rights, contracts, and obligations, Wilkins's language supplies simultaneously unambiguous signs to regulate those contractual exchanges. It frees the individual to pursue riches, pleasure, and credit within

23 It is the presence of this pragmatic marriage that best explains the divergent points of view on the essay presented by Slaughter and by Richard Kroll. What Slaughter finds in the essay is Aristotelian taxonomic essentialism, the drive to locate the nature of things themselves. What Kroll finds in the essay is "a highly sophisticated defense of the rhetorical or symbolic nature of human culture," the drive to authorize a more perfect language as a social construct (Richard W.F. Kroll, *The Material Word. Literate Culture in the Restoration and Early Eighteenth Century* [Baltimore/London: Johns Hopkins University Press, 1991], p. 186). If Slaughter and Kroll are both correct, both fail to see the conflict between these contradictory aims as central to understanding Wilkins's complex quest to authorize his discourse. For a fuller response to Kroll, see Robert E. Stillman, "Assessing the Revolution. Ideology, Language, and Rhetoric in the New Philosophy of Early Modern England," *The Eighteenth Century. Theory and Interpretation* 35, 2 (1994), 99–118.

a symbolic commerce that defines such interested pursuits, not by the despotic whims of Leviathans or the demotic opinions of the mob, but by "right reason" and law. By founding the authority of interest in common assent, and by defining interest (the pursuit of riches, pleasure, and credit) as a tool for perfecting natural virtues, Wilkins works to secure the promise from the perils of the Restoration's pervasive vocabulary of political economics.

Like all good contracts, Wilkins's essay fixes boundaries for essential matters demanding common assent, and allows latitude for less essential matters open to negotiation. His language supplies, for instance, the characters to assert that "a magistrate should have no power to reward or punish," but the meaning of *magistrate*, fixed by the Philosophical Tables, would signal immediately that statement's falsehood. (It would be no less obviously absurd to claim that maiz is a non-frumentaccious herb.) So too, it would be possible to state that "interest is the pursuit of infinite desires," but interest (as the Philosophical Tables make clear) is an instrument of virtue. Fools are never silenced, but folly can be graphically represented. On the other hand, nothing within the Tables would indicate whether Charles II acted rightly or wrongly in punishing Clarendon with his removal from office, or whether the Treaty of Dover was made in the national interest, or whether public monies should be spent to grow maiz. The meaning and value of particular historical actions are consigned, inside the Tables' negotiable boundaries, to the determinations of the best. Equipped with a philosophical linguistics, the best credit those actions according to universally applicable notions about magistrates and interests. Their very employment of the language would obligate them, under the terms of its linguistic contract, to esteem the truth and falsehood, the good and evil, of "things" according to its philosophical foundations.

Wilkins's essay begins, as I have shown, with a Prolegomena whose war against general custom establishes the need for a philosophical language. In the essay's second section, its "Scientifical part," the Philosophical Tables organize the universe of things into predictable relations, creating the foundation for the language's symbolic commerce and, hence, the principal focus for making its historical significance intelligible. However, in order to appreciate more fully, Wilkins's ingenuity in making his language inviting and engaging, it is also necessary to consider some of his most important strategies for mobilizing those tables into real characters and effable signs. Mobilization is the business of the essay's third or "*Organical* part": the description of the natural grammar, "that very Art by which Language should be regulated" (p. 19).

Wilkins's skill in constructing a natural grammar is most apparent in his provisions for a special category of signs called "Transcendental Particles." Displaying some pride in his ingenuity, he announces that these signs are "for the most part *new*" (p. 351).[24] Always aware of the need to economize, to create a language in which the number of substantive words (or "radicals," as they are sometimes called) would be *both* large enough to refer to all simple things and notions *and* small enough to be easily learned, Wilkins needed to devise some means by which the mere three thousand radicals of the Philosophical Tables could be combined to create more complex propositions. Transcendental particles are one of his chief solutions to this problem, designed as they are to "abbreviate the number of words, prevent much circumlocution, contribute to perspicuity and distinctness, and very much promote the elegance and significancy of speech" (pp. 319–320). They do not refer to things or notions, as the substantives of the Tables do; instead, similar to other particles, and to other parts of speech such as adverbs, pronouns, and interjections, they are tools for "circumstantiating" substantives, but in a more comprehensive, appropriately philosophical manner. As Wilkins writes, transcendental particles "do circumstantiate words in respect of some Metaphysical notion; either by enlarging the acception of them to some more general signification… [,] or denoting a relation to some other Predicament or Genus, under which they are not originally placed" (p. 318). Move from the characters of Wilkins's language to the Tables in which they are defined, and we enter a static universe of things hierarchically arranged by Genus. Add transcendental particles to those characters, and we can create logically determined pathways to relate the things of one Genus to those of another. They are called "transcendental" particles because—similar to the first three Genuses of the Tables, comprising "those Universal notions…belonging…properly to *Things*"—they function across the domain of the Tables to define universal relations of causality, instrumentality, quantity, quality, person, and action (p. 23). They are the grammar's principal means of mobilizing its metaphysical principles into language.

Nowhere is the power of Wilkins's language to illuminate the nature of things more apparent than in his account of the various transcendental particles. Consider, first, one example that he supplies among the metaphorical particles: Writing about a person who *blesses*, *delivers*, or *preserves* another, I employ a character indicating that these acts belong to the Genus of the Spiritual Actions of God, and I mark that character with a transcendental particle indicating, by metaphorical extension, that such actions are applied to a man. In this way, the language provides not merely a convenient tool for communicating, but also real knowledge about the "notions" communicated.

24 For an excellent study of the grammatical tradition out of which Wilkins's essay emerges, see G.A. Padley, *Grammatical Theory in Western Europe, 1500–1700* (Cambridge: Cambridge University Press, 1976).

Properly understood, *blessing*, *delivering*, and *preserving* have an innately divine significance, even in the human world. This is only one concrete example of the power of these particles to bring the transcendental values of the Tables to bear upon particular statements. Across the Genuses, the particles conveniently and economically mobilize more simple "things" into complex propositions, loaded with metaphysical meaning and value.

The transcendental particles are useful also in drawing attention, with a deliberateness and economy impossible in instituted languages, to how the world of things is constituted. Consider, again, a specific example: *Metal* is one of the substantives listed in the Philosophical Tables; *mine* (as in a gold "mine") is not listed. Writing about a mine, instead of using a distinct character, I add to the character for *metal* the transcendental particle of place, indicating, as I signify *mine*, that a mine is a place from which metals are extracted. The character supplies knowledge, then, about the real uses of things in the world. So too, if I wish to write about a kingdom (to use one of Wilkins's examples), I apply the same transcendental particle of place to the characters for *the king's authority*. By this means, I remind both myself and my reader, as I employ that character with its transcendental marker, about the real and essential nature of kingdoms: they are places in which kings have authority. The benefits of employing transcendental particles go far beyond convenience, economy, and elegance; they are also philosophical, and often patently political.

More simple natures are combined, upon an algebraic model, to yield complex propositions: *metal* + place = *mine*; *the king's authority* + place = *kingdom*. In order to employ Wilkins's language, again and again the user must perform similar calculations upon the simple, constitutive elements of reality. Describing Wilkins's language as a center of calculation is no mere figure of speech: calculating is the very essence, the organical part, of his natural grammar's effort to mobilize the universe of things and notions. Similar calculations upon the substantives of the Tables, signifying circumstances of causation, instrumentality, natural power, and time, yield similar benefits in drawing attention to the constitutive elements of reality. Add the particle for cause to *certain*, and I signify "to certify"; add the particle for instrument to *shaving*, and I signify a razor; the particle for power to *motion*, and I signify agility; and the particle for time to *nativity of Christ*, and I signify Christmas. *Certain* + cause = *certify*; *shaving* + instrument = *razor*; *motion* + power = *agility*; *nativity of Christ* + time = *Christmas*. In the center of calculation, the transcendental particles provide logical pathways for the permutation and combination of all things and notions, in such a way that the permutations and combinations constituting all real things are made evident in the very construction of the signs. A potential user of Wilkins's language does not set out to memorize its characters; rather, he labors first (for only a month, Wilkins assures us) to master the things and notions of the Philosophical Tables. Always, the language forces our attention back upon the world of things.

What is most revealing about the appeal of these transcendental particles that facilitate calculations upon the universe of things and notions is Wilkins's provision for various particles permitting the ready mobilization of credit or esteem. Creating transcendental particles to circumstantiate substantives according to excess or defect, perfection or corruption, augmentation or dimunition, permits the user of Wilkins's language to signify the "eminency of any thing in its kind" (p. 343). Add the augmentative particle to the character for *sound*, and I signify loudness; add that particle to *anger*, and I signify rage; add the particle to *fame*, and I signify renown. Couple the character for *moderation* with the diminutive particle, and I signify neutrality. As rage represents an augmentation of anger, neutrality represents a deficiency of moderation—good warning to anyone who might mistake Wilkins's moderation for neutrality.[25]

Beyond their convenience and economy, what is fascinating and significant about such signs is their power to create a fully commensurable universe of "things" in which objects, moral qualities, and social values are all subject to the same leveling principle of quantification. Once quantify the universe of "things," inside the twin domains of nature and culture, and make objects, morals, and values subject to algebraic calculations, then the consent of the best can mark everywhere, economically and unambiguously, the "eminency of any thing in its kind" (p. 343). Such calculations would be made, of course,

25 Just where did Wilkins stand on the great political issues of the 1660s? His most recent biographer argues that he was a moderate, occupying a neutral position outside and above the ideological fray; see Barbara Shapiro, *John Wilkins. An Intellectual Biography 1614–72* (Berkeley: University of California Press, 1969). *Moderation* is among Wilkins's favorite words, but to describe the complications of his public life as the expression of political "moderation" risks obscuring their meaning. Wilkins's moderation was not a sign of political neutrality. His opinions about sectarians such as Webster and republicans such as Harrington were fully political and ideologically motivated. Whatever the vagaries of political change, his profound conservatism in support of a strong centralized government founded upon traditional, patriarchal values was unabating, from his fierce opposition to the horrors of the civil wars in his *Discourse Concerning the Beauty of Providence* (1649) to his unrelenting service on behalf of royal causes as Charles's most loyal clerical spokesman in the House of Lords during the 1660s. More tool than doctrine, *moderation* was the politics of interest in action—the vehicle for a commitment to philosophical modesty, providential faith, and political quietism, allied as a trinity of responses to his lived experience amidst the cultural crises of the Interregnum. A new biographical treatment of Wilkins is sorely needed.

not inside the restricted boundaries of a particular science, like Graunt's statistics, or even within the more comprehensive political vocabulary enjoined as authoritative by a Hobbesian sovereign—rather, they would occur in every utterance on every occasion of public commerce. No wonder the language's benefits are deemed universal. By excluding from the Philosophical Tables customary titles of honor and office (signs for Duke or Earl, Knight of the Garter or President of the Royal Society), Wilkins asks his sovereign elites to relinquish to the realm of the merely historical their customary signs of power and position. In compensation, his transcendental particles signify degree and eminence of a different, essential, and universal kind. To be a Baron is to enjoy a particular title in a particular historical world. To acquire a character for eminence in the universal language (of office, virtue, or renown) is to be eminent inside a universal order whose rightness is confirmed by the consent of the best and the nature of things.

It would be too long a task to describe all of Wilkins's strategies for mobilizing the Tables' accumulation of things, and a redundant exercise to describe the technical vehicles that he creates to effect that mobilization, the language's real character and effable signs. Instead, as a means of both illustrating and extending arguments already advanced, it is helpful to examine briefly one of his major proofs of his language's facility and usefulness, his transposition of the Lord's Prayer (see Figure 5.1) and the Apostle's Creed into the real character. *Defamiliarizing* is perhaps too mild a word to describe the experience of first confronting Wilkins's characters. In place of the familiar Bible words, we find startlingly new characters, composed of lines, hooks, circles, and points. At first sight, the real characters more closely resemble mathematical symbols than words. (His effable signs are only slightly less unfamiliar, more reminiscent of Swift's renderings of Lilliputian than King James English.) But once allow the unfamiliarity of these characters to diminish, and their ingenuity is impressive: conspicuously economical, they require only slightly more than three short lines to render the Lord's Prayer that English represents in six fuller lines.

Move from the real characters to Wilkins's annotations, explaining for the reader's convenience how he arrived at each symbol, and the language's massive ingenuity seems still more impressive. Wilkins's characters do not attempt to translate the Lord's Prayer or the Apostle's Creed word for word. As is appropriate for a philosophical language, its characters represent those prayers notion by notion, drawing attention, by the very nature of its signs, to substantive "things" in the scriptural text and clearly discriminating between those substantives and the signs meant to circumstantiate them.

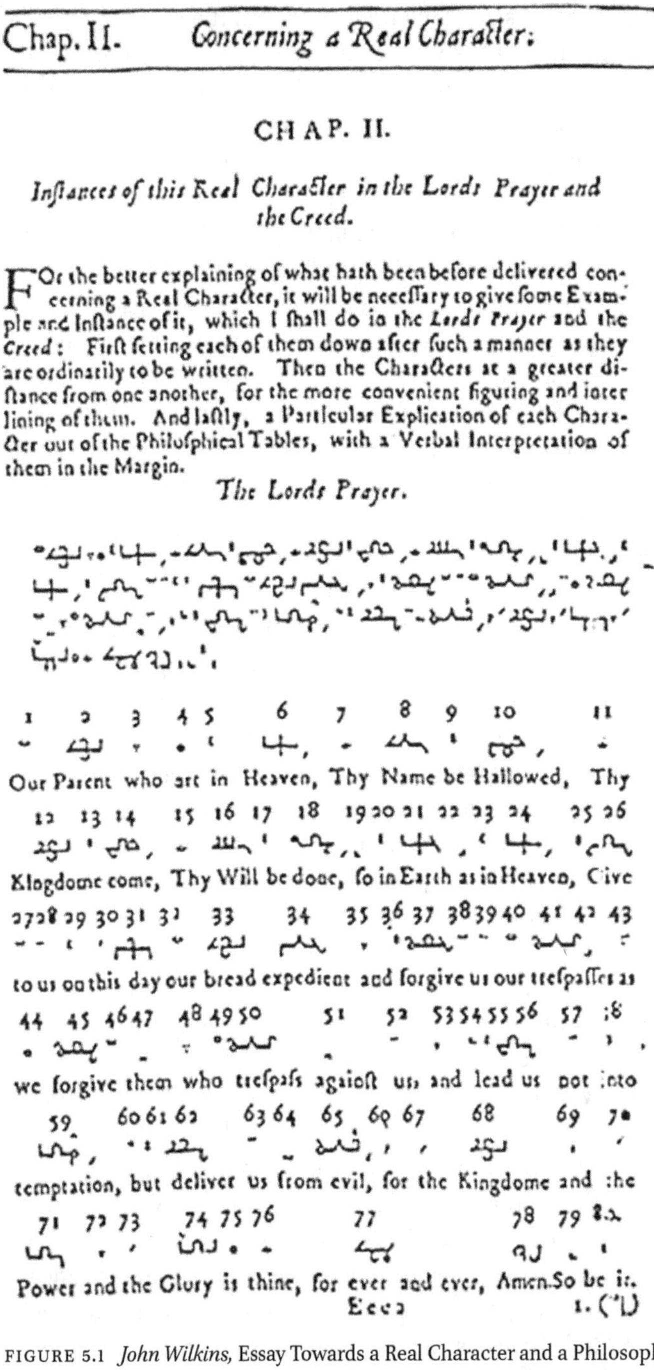
Chap. II. *Concerning a Real Character.* 395

CHAP. II.

Instances of this Real Character in the Lords Prayer and the Creed.

FOr the better explaining of what hath been before delivered concerning a Real Character, it will be necessary to give some Example and Instance of it, which I shall do in the *Lords Prayer* and the *Creed*: First setting each of them down after such a manner as they are ordinarily to be written. Then the Characters at a greater distance from one another, for the more convenient figuring and inter lining of them. And lastly, a Particular Explication of each Character out of the Philosophical Tables, with a Verbal Interpretation of them in the Margin.

The Lords Prayer.

1 2 3 4 5 6 7 8 9 10 11
Our Parent who art in Heaven, Thy Name be Hallowed, Thy
12 13 14 15 16 17 18 19 20 21 22 23 24 25 26
Kingdome come, Thy Will be done, so in Earth as in Heaven, Give
27 28 29 30 31 32 33 34 35 36 37 38 39 40 41 42 43
to us on this day our bread expedient and forgive us our trespasses as
44 45 46 47 48 49 50 51 52 53 54 55 56 57 58
we forgive them who trespass against us, and lead us not into
59 60 61 62 63 64 65 66 67 68 69 70
temptation, but deliver us from evil, for the Kingdome and the
71 72 73 74 75 76 77 78 79 80
Power and the Glory is thine, for ever and ever, Amen. So be it.
Eee 2 1.

FIGURE 5.1 *John Wilkins,* Essay Towards a Real Character and a Philosophical Language *(London, 1668), p. 395: the Lord's Prayer written in the real character*
[EARLY ENGLISH BOOKS ONLINE, ACCESSED 22 APRIL, 2014.]

As always, Wilkins's language strives relentlessly to turn the reader's attention from words to things, stripping the verbal texture from the scriptural text to illuminate through its transparent characters the Logos as reason, not word.

Each of Wilkins's substantive characters forms a semantic bundle out of which the reader must unpack a whole complex of knowledge. The characters refer to things not words. Organized as a kind of philosophical zip code, each sends the reader directly, by its particular configuration of lines, hooks, circles, and points, to a "simple" thing or notion in the Philosophical Tables, appropriately circumstantiated and mobilized by additional markers inside the sentence into a complex proposition. (Neatly enough, each item belonging to a particular Genus shares the same character, consistently variable to specify a particular Difference and Species; related things achieve, thereby, related characters.) What is especially impressive about Wilkins's scheme, beyond its logic, convenience, and economy, is not just its adequacy to express a universe of things and notions, but also its power to regulate the semantic boundaries inside which these characters negotiate meaning and value.

Consider the appeal of negotiating semantic boundaries. Meeting the phrase "forgive us our trespasses" in the English of the Bible, readers will understand the word *trespasses* more or less precisely according to the conventions of a particular interpretive community (a church, prayer group, literary club, etc.), or according to the individuating (and sometimes idiosyncratic) circumstances of their own family, education, religious training, or purely personal experience; its range of possible meanings is in principle limitless, and in practice, limited only by history's vast range of diverse communities, value systems, and idiosyncratic readers. But rendered into a real character, *trespasses* is regulated within far more precise boundaries, which provide real knowledge about both the essential nature of trespasses and how they are committed. Included within the Genus of Transcendental Generals, within "the more *common and absolute Differences of things*," *trespasses* belongs to the Species "which doth include a respect to the Will, as to their agreement or disagreement with that Faculty, whereby they are rendered desirable, or avoidable; which we call *Goodness* or *Evilness*" (p. 400). Linked to the will and its always dangerous desires, trespasses are evil actions for which individuals must take responsibility, since, in kind, they are avoidable. Linked to other absolute differences among things, from falsehood and spuriousness to mixedness and imperfection, trespasses assume a place among the comprehensive range of actions in nature and society, from promiscuous compoundings to bastard productions, whose kind indicates heterogeneity, most often corruptive; as such, they gain much of the concreteness, the material reality, of those "things" with which they are associated. Linked, too, to Transcendental Generals, trespasses possess a universal, unchanging essence; to commit a trespass is not to violate a historically constructed or customary norm, but (properly understood) to perform an action whose essence participates in universal and unchanging evilness itself.

Wilkins's semantic boundaries do not permit the reader to know whether any particular action constitutes a trespass. What his language does provide is a framework in which to understand the nature of trespasses, and provisions within that framework for the consent of the best to credit the reality of any putative act of trespassing. Under the terms of Wilkins's linguistic contract, obligating the user to confer meaning and value by the provisions of the Tables, the potentially limitless interpretations of the Biblical phrase "forgive us our trespasses" are safely regulated inside semantic boundaries that fix its meaning within a traditional metaphysics and permit negotiations about its specific applicability according to the Restoration elites' patriarchal code of values. In a process universally extendable, regulating *trespasses* in language makes possible the regulation of trespasses in history. Mastering history by mastering language achieves, by this means, a new technological precision.

Illuminating the language's power to regulate semantic boundaries is most easily accomplished by examining what happens to some of the Restoration's most politically charged words as they are transposed from the Bible's English into Wilkins's real character. When read aloud from the pulpit of any Anglican church in the 1660s, the Creed's reference to "the Communion of Saints" would have carried potential reverberations—not just with the sainted members of church history, but also with those fearsome Saints who followed Cromwell into battle during the civil wars. Safely contained within the Genus of Ecclesiastical Relations, among the "species denoting such as are *eminently religious*," the Saints are philosophically purged from any uncomfortable or distracting historical associations (p. 412). Empty of history, the characters strive for universal significance.

Semantic boundaries are regulated, too, as Wilkins disposes of potentially troublesome metaphors. The prayer for "daily bread" is transposed into a real character that defines *bread* in a curiously literal manner: under the Genus of Oeconomical Provisions, in the Difference of "Sustenation ordinary," the character for bread refers to that kind of food "made of Grain" (p. 399). Immediately diminished are all of the word's latent metaphorical meanings, signifying wide varieties of spiritual nourishment or specific sorts of economic liberties desired by leveling fanatics who demand "daily bread" of a seditiously political kind. Nothing prevents the user of Wilkins's language from adding a transcendental particle of metaphor to the character for bread—but such a particle, again, could be accredited, properly, only by the consent of the best.

Everywhere the business of defining assumes the form of individual acts of calculation. The potential user of Wilkins's language is constantly instructed by the annotations to the Lord's Prayer and the Apostle's Creed in how to make the necessary permutations and combinations to build the simple "things" of the Tables into complex propositions. In order to express a complex notion, such as temptation, users must know the simple, constitutive notions from which temptation is formed, and the logical pathway that leads from those notions to the real state of temptation. They must know, therefore (or *learn* in the struggle to use the language), that temptation is best defined among the "General Relations of Actions Comparate," as a specific variety of the "*Trying*, or the Examining of things, for the distinguishing of their Truth and Goodness" (p. 401). They must consider, further, that superimposed upon this simple notion of trying or examining is the notion of hurtfulness, a circumstantiating quality added (quite literally) to it, as the pathway that leads from morally neutral examinations to morally reprehensible temptations. Calculating how to construct an appropriate character for temptation, they must add to the sign denoting the Genus of Transcendental Relations of Action specific markers for the act of comparing, and add, once more, the transcendental particle of corruptive to impart appropriate moral force to that action. Definition proceeds, then, by extended and often extensive calculations across the domains of nature and culture. Designed to compel its readers to make such calculations, Wilkins's language operates as an educational primer for the speculative understanding.

Sometimes the business of calculating takes place simply and directly: add the transcendental mark of person to the character for *creation*, and the sign refers to maker or creator; add the transcendental mark of place to the pronoun *which*, and the sign refers to "the place in which." But once add the transcendental particle of augmentation to the character for *reputation* at the conclusion of the Lord's Prayer, and something altogether more complex and interesting transpires: in a single moment of calculation, as the particle is superimposed, reputation is transformed into Glory, "the greatest kind and degree of Reputation"—as both one more example of the work of the speculative understanding in calculating to define, and an instance of that final end in which all acts of calculation achieve their limit and objective, the credit due to divine glory itself (p. 402). At the moment that the philosophical language esteems divine glory, it exemplifies through its own real characters and natural grammar the power of calculating among simple, transparently disclosed things to rescue logic from the Logos, reason from the Word. In making that rescue, Wilkins's language provides not just a justification for the Society's recently institutionalized claims upon the public interest, but also a new technological vehicle, fueled by the best new philosophical learning, to secure the universal interest.

CHAPTER 6

"The Spirit of Invention"

Hooke's Poetics for a New Science in An Attempt to Prove the Motion of the Earth by Observation

Frédérique Aït-Touati

An Attempt to Prove the Motion of the Earth by Observation[1] was the first of Robert Hooke's famous Cutlerian Lectures, delivered in 1670 and published in 1674. It is an account by the experimentalist of his work following the path of a star in the head of the constellation of Draco. The stellar parallax he could see and measure was the proof, said Hooke, of the motion of the earth around the sun. Though the last page of this *Attempt* is frequently alluded to,[2] as it formulates the famous "System of the World" that Newton read and completed, the rest of the text has received little attention.

My concern here will not be these famous pages of the *Attempt*, but rather, the rest of the text and its preface. When the preface and the *Attempt* are considered together, what appears is a pervading "Spirit of Invention," in Hooke's own words. This Spirit of Invention relates to the text and its projects in two ways: first, the invention and construction of instruments that can prove the Copernican Hypothesis; second, the conception and construction of texts that can faithfully convey the discoveries, both of art and nature. I would like to show how these mechanical and poetical programmes were thoroughly intertwined in Hooke's conception of and plea for a new science and a new method.[3] What

1 Robert Hooke, *An Attempt to Prove the Motion of the Earth*, London, Printed by T.R. for John Martyn, Printer to the Royal Society at the Bell in St. Pauls Church-yard, 1674. I would like to thank Jonathan Morton for his careful reading and help with the English.

2 See for example Alexandre Koyré, "An Unpublished Letter of Robert Hooke to Isaac Newton," *ISIS* 43, 4 (1952), 312–337, and Ofer Gal, *Meanest Foundations and Nobler Superstructures. Hooke, Newton and the "Compounding of the Celestiall Motions of the Planetts"* (Kluwer, 2002), Chapter I, 20.

3 My study is thus in line with John Harwood's pioneering work on Hooke's rhetoric. Harwood showed that the specific rhetoric of *Micrographia* should be understood in the context of the "identifiable social and political concerns of the early Royal Society. The early Royal Society simultaneously pursued two activities, one scientific and the other rhetorical: 'doing' the New Philosophy and 'writing' about it." In this context, printing was "a way to establish, enhance, and protect its public image." John T. Harwood, "Rhetoric and Graphics in Micrographia," in *Robert Hooke. New Studies*, eds. Michael Hunter and Simon Schaffer (Woodbridge, Suffolk, Boydell, 1989), 131.

 | DOI 10.1163/9789004283701_008

are the literary strategies by which Hooke tried to convey both his results and his method? To what extent did he use or disregard the existing conventions of scientific writing? I shall try to answer these questions in relation to the general theme of this journal's issue: was Hooke still writing in the age of Bacon? I shall answer that insofar as one reads Bacon the way Hooke did, he definitely was.

The preface has generally been read as a text written in 1679,[4] the date of the publication of the *Cutlerian Lectures* as a whole. The two original copies of the *Attempt* that still exist in the Wren Library at Trinity College, Cambridge, demonstrate that the preface is in fact a text from 1674,[5] published with the *Attempt*.[6] With this revised dating in mind, I shall analyse what Hooke himself presented as a programme—not only a programme of research, but also a method, and indeed a poetics suited to the new science. In studies of Hooke, it has been common to deplore the incomplete aspect of his work. I shall try to reconsider this fragmentary aspect in the light of Hooke's poetics, and shall suggest that the style he adopted in his various writings was in keeping with his method of inquiry. For the sake of clarity, I will follow the three main parts of rhetoric that can be recognised in Hooke's preface: *inventio*, *dispositio* and *elocutio*. However, I will mostly be concerned with questions of poetics and I will try to show that these categories[7] (invention, arrangement and style) are intended to shape a new scientific genre.

---

4 Koyré, Alexandre, *op. cit.*, 319.

5 The traditional evaluation of Hooke is recalled, for instance, by Koyré, *op. cit.*, 314, n16: Hooke "totally lacked Newton's powers of concentration. His mind was restless, continually disturbed by fresh ideas, but they were nearly all good, and many were of first importance." And 319: "Hooke certainly is perfectly right in insisting on his priority. Yet it cannot be denied that the lacuna which we discovered in his earlier work has not been filled: Hooke still does not know, 'what the several degrees are' by which the attractive power varies with the distance. In 1678, when he published his *Cometa*, he is as far from the solution of that problem as in 1674 and that is probably why, feeling that he is unable to keep his promise and to 'explain' his 'system of the world,' he simply reissues, in 1679, his old *Attempt* under the new cover of *Lectiones Cutlerianae*." In my opinion, the publication of the *Lectiones* is not a "new cover" but rather the realisation of Hooke's precise programme of publishing as announced in 1674: to collate his discoveries in one volume. Again, 320, n. 45, Koyré quotes a long passage from the preface (dating it from 1679) with the following comment. "Yet, it was this very restlessness, the inability of concentration, and therefore, of obtaining conclusive results, that made him unacceptable to Newton." Newton, according to Professor Pelseneer, was a "classical" mind and must have shuddered when reading Hooke's "profession de foi."

6 See also the bibliography by Keynes which notes this preface in the edition of 1674. Geoffrey Keynes, *A Bibliography of Dr. Robert Hooke*, Oxford, 1960, 30.

7 For Hooke "invention" should be understood broadly in its rhetorical sense. Indeed, it is taken in this broad sense by the various *ars poetica* written in the same period: *inventio*, the first part of rhetoric, generally encompasses the two others, *dispositio* and *enunciatio*.

Inventio

Inventio (invention) was the first of the five parts of the classical art of rhetoric.[8] In Hooke's preface, however, this central rhetorical category takes on a new meaning.

> The greatest part of Invention being but a lucky hitt of chance, for the most part not in our own power, and like the wind, the Spirit of Invention bloweth where and when it listeth, and we scarce know whence it came, and whither 'tis gone…'twill be much better therefore to imbrace the influences of Providence, and to be diligent in the inquiry of every thing we meet with.[9]

If this sounds like a rather surprising programme of research, it will help to recall what Bacon himself had defined as "invention" in the *Advancement of Learning* in 1605. In the first of the two paragraphs quoted below, Bacon distinguishes between scientific and rhetorical inventions. In the second, he explains why he thinks that scientific invention should be different from traditional rhetoric invention:

> XIII. (1) Invention is of two kinds much differing—the one of arts and sciences, and the other of speech and arguments. The former of these I do report deficient; which seemeth to me to be such a deficience as if, in the making of an inventory touching the state of a defunct, it should be set down that there is no ready money. For as money will fetch all other commodities, so this knowledge is that which should purchase all the rest. And like as the West Indies had never been discovered if the use of the mariner's needle had not been first discovered, though the one be vast regions, and the other a small motion; so it cannot be found strange if sciences be no further discovered, if the art itself of invention and discovery hath been passed over.
>
> (6) The invention of speech or argument is not properly an invention; for to invent is to discover that we know not, and not to recover or resummon that which we already know; and the use of this invention is no other but, out of the knowledge whereof our mind is already possessed to draw forth or call before us that which may be pertinent to the purpose which we take into our consideration. So as to speak truly, it is no invention, but a remembrance or suggestion, with an application; which is the cause why the schools do place it after judgment, as subsequent and not

8 *Inventio, Dispositio, Elocutio, Memoria, Actio.* Invention, Arrangement, Style, Memory, Delivery.

9 Hooke, *An Attempt*, *op. cit.*, "To the Reader" (unpaginated).

> precedent. Nevertheless, because we do account it a chase as well of deer in an enclosed park as in a forest at large, and that it hath already obtained the name, let it be called invention; so as it be perceived and discerned, that the scope and end of this invention is readiness and present use of our knowledge, and not addition or amplification thereof.[10]

With Bacon, invention becomes the condition of the progress of science. Here as often, he took an old concept and gave it a new meaning. Bacon's notion of invention is a method designed to discover a hitherto unknown nature, completely removed from the notion of imitation found in classical rhetoric. It was this Baconian programme that Hooke presented in a fresh form for his Restoration audience. Like Bacon, Hooke attached a special significance to what could be learnt about nature, not just through passive observation, but also from experimental manipulation. Invention, then, should be construed in the active sense of Bacon's *scientia operativa*: an "operational science." In the epistemology of the "maker's knowledge," the process of invention is always an active posture, in which art and nature constantly interact.[11] In Hooke as in Bacon, art and nature were deeply interwoven. Where Hooke went far beyond Bacon, however, was in his insistence on the role of instruments, as Jim Bennett has shown. Hooke's notion of invention was very close to the notion of ingenious technique. In his Lecture of 1678, entitled *Lampas*, Hooke explicitly linked Invention and Industry. In the *Attempt*, he demonstrated in dramatic fashion the efficiency of his mechanical Inventions (Figure 6.1).

This plate, printed alongside the text, combines diagrams of the parallax and drawings of the optical and measuring instruments that Hooke designed for his observation: various micrometers, and most importantly, a zenith telescope. In the centre a large telescope is drawn, mounted through the roof of Hooke's rooms in Gresham College and combined with a tube and various additional lenses. This formed a complex chain, a vertical laboratory pointing toward the sky. Thanks to this powerful telescope, the experimentalist was capable of following with precision the path of the star that was projected onto the limited and controlled space that he had marked with graduations.[12]

10 Francis Bacon, *The Advancement of Learning*, ed. G.W. Kitchin London: (Everyman's Library, sixth edition, 1984), 127.

11 Antonio Pérez-Ramos, *Francis Bacon's Idea of Science and the Maker's Knowledge Tradition* (Oxford: Oxford University Press, 1988).

12 Hooke's device is perfectly in keeping with the perspective machines and optical devices that had been used by painters and natural philosophers alike since the Renaissance. See Martin Kemp, *The Science of Art. Optical Themes in Western Art from Brunelleschi to Seurat* (New Heaven and London: Yale University Press, 1990), Chapter IV: "Machines and Marvels."

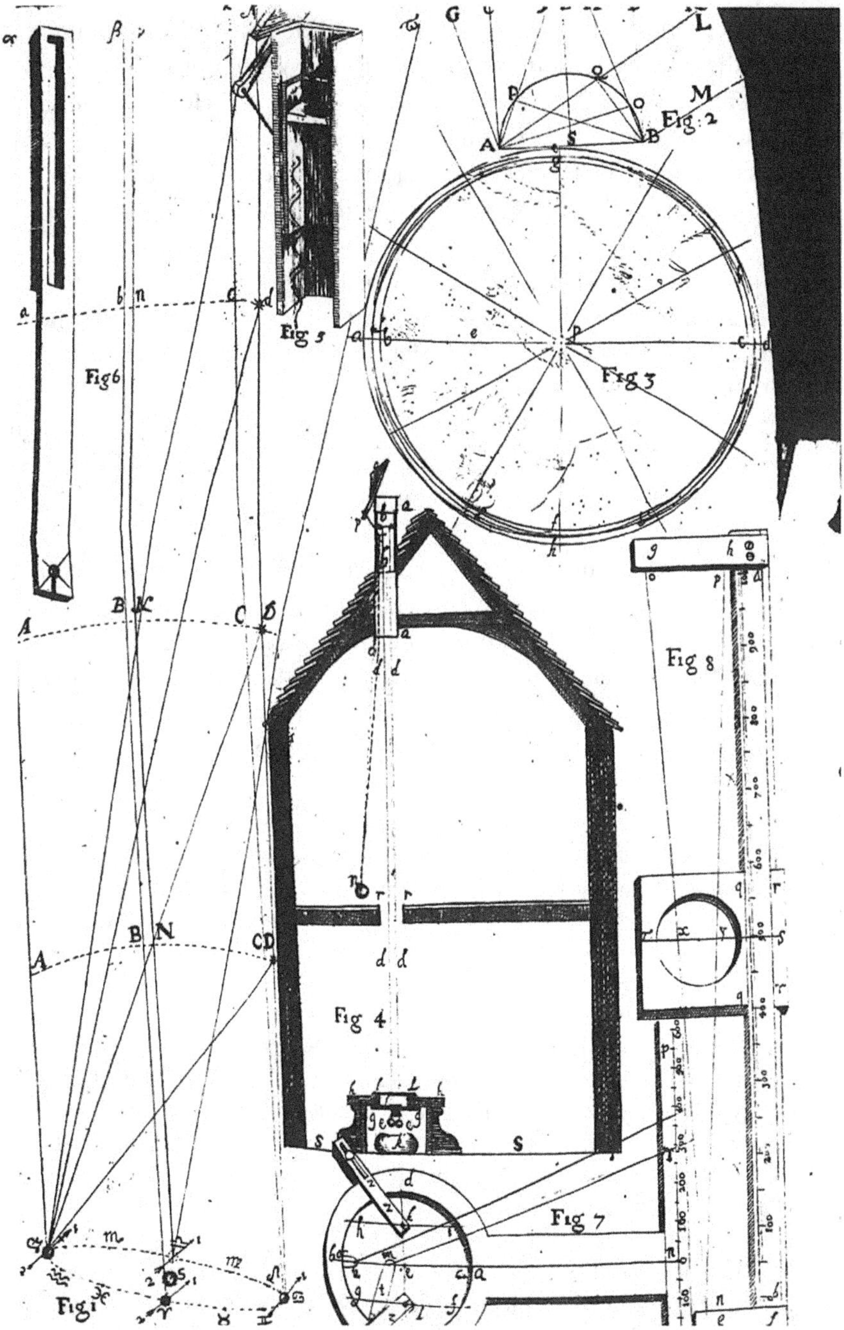

FIGURE 6.1 *Plate showing various micrometers and a zenith telescope from Robert Hooke,* An Attempt to Prove the Motion of the Earth by Observation *(London, 1674).*
[EARLY ENGLISH BOOKS ONLINE, ACCESSED 29 OCTOBER, 2014]

As a "mechanic,"[13] Hooke was very aware of the practical and commercial value of these inventions. In fact, at that time invention was becoming the principal rationale for the establishment of a patent, and Hooke would sometimes have to battle with his numerous competitors to obtain such patents. The preface reflects this concern very clearly. Hooke's construction of a new scientific genre was also, therefore, a means to protect his inventions:

> There are other conveniences also in this Method of Communication not less considerable then the former, amongst the rest the securing of Inventions to their first Authors, which 'tis hardly possible to do by any other means; for there are a sort of Persons that make it their business to pump and spy out other Inventions, that they may vend them to Traders of that kind, who think they do ingenuously to print them for their own, since they have bought and paid for them.[14]

When one compares one of the first legal texts stating the word "invention" in the context of patent law, the Statutes of Monopolies (1623),[15] and Hooke's proposal for "the securing of Inventions to their first Authors," it is clear that Hooke was calling for a new kind of text capable of both disseminating and protecting knowledge. The importance given to the commodification and the protection of knowledge was in keeping with Hooke's persona as a mechanic, as explained by Steven Shapin.[16]

---

13 On this question and its role in Hooke's ethos, see Jim Bennett, "Robert Hooke as Mechanic and Natural Philosopher" in *Notes and Records of the Royal Society of London*, vol. 35 (1980), 33–48; Jim Bennett, "Hooke's Instruments for Astronomy and Navigation," in *Robert Hooke. New Studies, op. cit.*, 21–32, and Steven Shapin, "Who Was Robert Hooke?" in *Robert Hooke. New Studies, op. cit.*, 253–285.

14 Hooke, *An Attempt, op. cit.*

15 "VI. Proviso for future Patents for 14 Years or less, for new Inventions 'Provided alsoe That any Declaracion before mencioned shall not extend to any tres Patents and Graunt of Privilege for the tearme of fowerteene yeares or under, hereafter to be made of the sole working or makinge of any manner of new Manufactures within this Realme, to the true and first Inventor and Inventors of such Manufactures, which others at the tyme of makinge such tres Patents and Graunts shall not use, soe as alsoe they be not contrary to the Lawe nor mischievous to the State....'"

16 "...if the pattern of Hooke's behaviour was depreciable in a Christian gentleman, it was widely considered to be nothing exceptional among tradesmen. Indeed, contemporary social guides to the code of English gentility stressed the contrast between the openness, the generosity and the reliable truth-telling of the gentleman and the secretiveness, the 'sordid interests' and the duplicity of the tradesman and merchant," Shapin, Steven, "Who Was Robert Hooke?" in *Robert Hooke. New Studies, op. cit.*, 276.

What Hooke meant by his phrase "the Spirit of Invention" becomes still clearer if one remembers the context of the debate with Henry More over the "Spirit of Nature," a debate which was revived by the publication in 1671 of More's *Enrichidion Metaphysicum* which defines the "Spirit of Nature" as follows:

> The Spirit of Nature—is A Substance incorporeal, but without Sense and Animadversion, pervading the whole matter of the universe, and exercising a Plastical power therein according to sundry predispositions and occasions in the parts it works upon, raising such Phaenomena in the World as cannot be resolved into mere mechanical powers.[17]

More's main point is that there is no purely mechanical principle in nature. He was defending the "immaterial spirits" against Cartesian and Baconian mechanism that he considered as contradicting religion.[18] With the expression "Spirit of Invention," Hooke was cunningly reversing More's concept. In Hooke's text, indeed, the notion of invention, as we have seen, is deeply mechanistic. If there is indeed a little dig against More, it is especially witty in the provocative oxymoron of the terms "spirit" and "invention."

It should be noted, finally, that the sentence mentioning the "Spirit of Invention" ("and like the wind, the Spirit of Invention bloweth where and when it listeth, and we scarce know whence it came, and whither 'tis gone"[19]) is a direct quotation of the Gospel of John: "The wind bloweth where it listeth, and thou hearest the sound thereof, but canst not tell whence it cometh, and whither it goeth: so is every one that is born of the Spirit" (King James Version, 3:8).

This is in Chapter 3 of the Gospel of John, a chapter mostly concerned with the Spirit of God. Bearing that in mind, the "Spirit of Invention" and the "Providence" which is alluded to shortly afterwards take on a new meaning. Hooke believed deeply in the serendipity of discovery. He thus came to distinguish two manners of inquiry: by chance, and by Design. His commitment to the first, chance against Design, attests his well-known optimism into the capacity of mankind to find ceaseless novelties in the wonders of nature and art.

---

17 More, Book III, Chap. 12.

18 See for instance Robert A. Greene, "Henry More and Robert Boyle on the Spirit of Nature," in *Journal of the History of Ideas* 23, 4 (Oct.–Dec. 1962), 451–474.

19 Hooke, *An Attempt, op. cit.*, "To the Reader."

Dispositio

Hooke's endeavour in natural philosophy would be achieved, he argued, through a particular method for ordering the findings. This is the second aspect of Hooke's poetics: the arrangement. In fact, Hooke professed a rather unorthodox conception of *dispositio*:

> I have complied, with the desire of several of my Friends (though otherwise not thereunto obliged) to commit divers of those Discourses to the Publick, though of themselves for the most part incompleat, and Essayes or Attempts only upon several Subjects which have no dependencie or coherencie one with another.[20]

Beyond the usual modesty of the preface, Hooke was in fact advocating his method of writing as that most suited to natural philosophy. Such method can be summarized in few principles: a refusal of any preconceived order, the voluntary incompleteness of findings, and their variety. The existing way of writing science indeed suffered from a series of inconveniences:

> In the doing hereof, I design to avoid any kind of Method or Order that may require Apologies, Prefaces, or needless Repetitions of what is already known, or might have been said upon that Occasion, or may necessitate me to follow this or that Subject, that doth not some way or other offer it self as it were, and prompt me the consideration thereof.[21]

The traditional scholarly composition of the treatise was imposing an artificial structure upon what Hooke considered as the natural random order of discovery. One should follow all the topics which might arise at any point rather than systematically undertaking to discover everything about one. Hooke advocated a mode of writing in keeping with this pragmatic method: only the useful, the new and the necessary elements of the demonstration are retained. Hooke calls for a reformation in writing which is as radical as the "Reformation of Philosophy" he was promoting in the Preface of *Micrographia* in 1665.[22]

---

20 Robert Hooke, "To the Reader," *op. cit.*

21 *Ibid.*

22 Interestingly, the statutes of the Royal Society published in 1728 seem to build from Hooke's rules of writing that I have just quoted: "In all Reports of Experiments to be brought into the Society, the Matter of Fact shall be barely stated, without any Prefaces,

This method was fully in line with the Baconian collaborative ideal, in which each member of the philosophical community should build on the findings of other members. Such collaboration supposed intense communication and increased exchanges, in which texts would play a crucial role. What follows therefore is not only a poetics but a programme of publication:

> But because they may possibly admit of some better order hereafter, I design to print them all of the same Volume, that so they may be, when ranged, either stitched or bound together, and may, as occasion requires, be referred to under the Title of their Number and Page. This way I chuse as the best for promoting the Design of this Lecture; for as there is scarce one Subject of millions that may be pitched upon, but to write an exact and compleat History thereof, would require the whole time and attention of a mans life, and some thousands of Inquiry, whatever it be....[23]

The rest of the *Attempt* manifests the same defiance against any kind of *a priori* order that might be applied to the universe, in particular the notion of harmony that he deeply criticises as insufficient as a scientific proof. There is therefore a strong correlation between the chosen experimental protocol and the publishing technique of a series of independent tracts. The inevitable incompleteness of each history was seen not as a defect but rather as a consequence of a wandering way of inquiry. Such a conception of *dispositio* might well explain the paratactic structure typical of Hooke's writings, whether in *Micrographia*, the *Cutlerian Lectures* or the *Posthumous Works*.

Indeed, Hooke's poetics had already been beautifully illustrated in his *Micrographia*, his famous treatise on microscopy published in 1665. From the infinitely small to the infinitely remote, *Micrographia* encompassed the complete scale of nature and displayed it through ingenious composition.

In the final plate and observation from *Micrographia*, supposedly added to fill a "void," Hooke demonstrated his mastery of optical technology: "Having a pretty large corner of the Plate for the seven Starrs, void, for the filling it up, I have added one small Specimen of the appearance of the parts of the Moon, by describing a small spot of it...." (Figure 6.2)[24]

Apologies, or Rhetorical Flourishes, and entered so into the Register-Book, by order of the Society" (Quoted in RFJ, 84 and Vickers, 14).

23 Robert Hooke, "To the Reader," *op. cit.*

24 Robert Hooke, *op. cit.*, "Observ. LX. Of the Moon," 242.

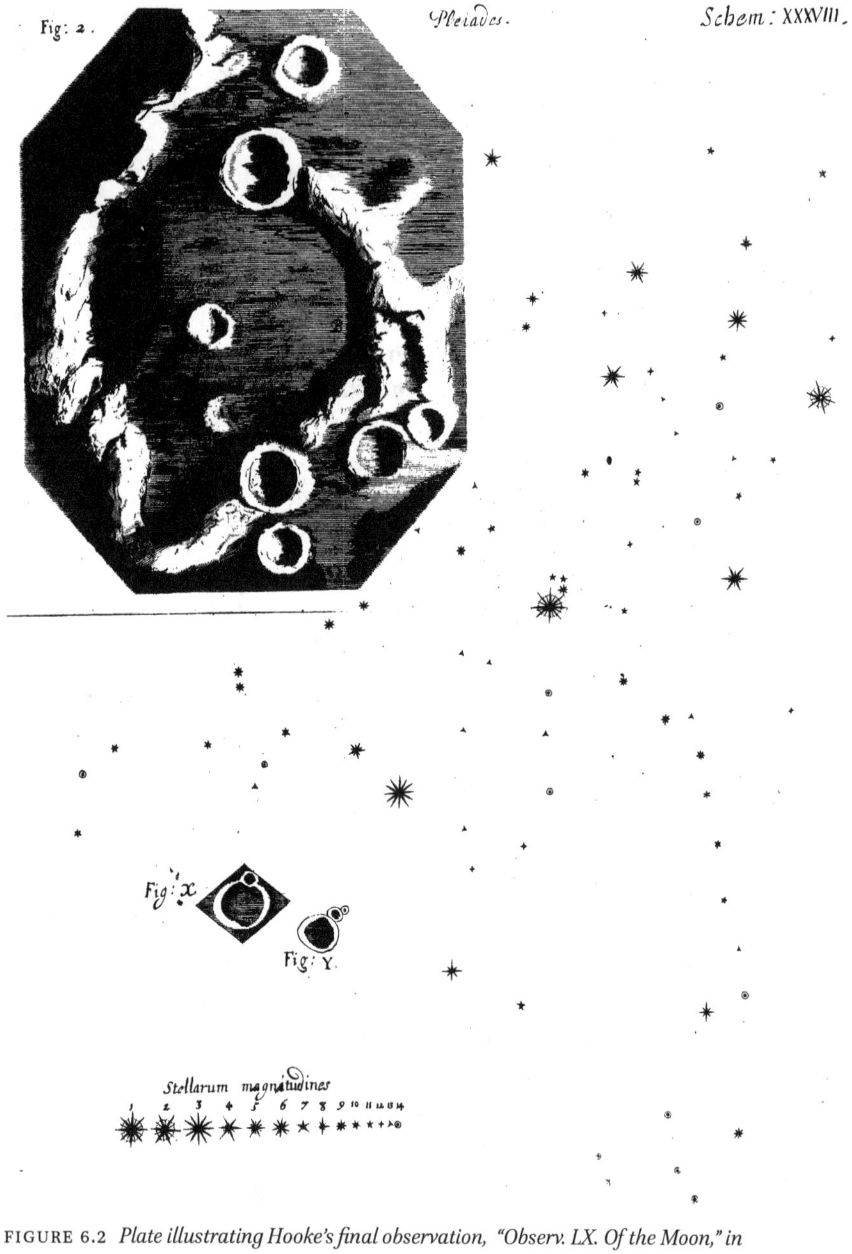

FIGURE 6.2 *Plate illustrating Hooke's final observation, "Observ. LX. Of the Moon," in* Micrographia *(London, 1665)*
[SOURCE: EARLY ENGLISH BOOKS ONLINE, ACCESSED 29 OCTOBER, 2014]

Adding a word or thought to a text that is already complete is a rhetorical technique to draw emphasis to the addition. Hooke's final observation is a perfect hyperbaton. This graphic trope is, however, more than pictorial wit. It is significant, I believe, of Hooke's general method: an accumulative method which does not set any *a priori* end to the process of discovery. The poetics of Hooke's scientific writing is deeply paratactic, and could be construed as the presentation of a series of independent elements to be juxtaposed. Hooke was making eclectic rather than exhaustive collections.

His method of research as well as his method of writing, then, was very much based on the notion of sampling, a term that Michael Hunter[25] and Nick Wilding[26] have used to designate various texts by Hooke. In that sense, I would suggest, Hooke's works might be understood as one vast store-house. All his texts together form a vast collection of samples of nature. It might be useful to recall at that stage that the metaphor of the "store-house"—first used by Bacon—was precisely the one used by Hooke to explain the functioning of memory.[27] In another context, the image ties in with Hooke's role of organising scientific displays for the public, being responsible for the Royal Society's Repository, that is, the Royal Society's museum and cabinet of curiosities. Here, we find a striking link between the question of rhetoric *dispositio*, a new kind of scientific genre, allowing for the accumulation of pieces of knowledge, and the very material aspect of the organisation of knowledge in space.

---

25 "Again, there were Baconian precedents, in terms of the queries that Bacon had prefixed to each of the 'Histories' that he had seen as exemplifying the method of his Great Instauration. But Hooke went beyond this, particularly in the implication that the lists that he provided in the General Scheme were merely a sample of such lists that it would be an easy matter to produce for all the other phenomena of the universe. Indeed, at least in potential, he went beyond his mentor, Boyle, who followed Bacon's example in producing a more limited range of lists of 'heads' of distinct topics. In this regard, Hooke has claims to be considered one of the most systematic exponents of scientific method of his age" (my emphasis). Jim Bennett, Michael Cooper, Michael Hunter and Lisa Jardine, *London's Leonardo. The Life and Work of Robert Hooke* (Oxford, Oxford University Press, 2003).

26 "The lecture [on memory] repeatedly refers to the memory as a 'repository,' 'storehouse' and so on. In its non-representational concern with 'things' not 'words,' memory's nature, use and problems are identical to those of an actual repository or museum. These were not abstract problems for Hooke in the late 1670s: issues of the correct selection of objects, organization, access to an use of the Royal Society's Repository were a constant concern in Hooke's efforts to transform it from an aristocratic cabinet of curiosity (producing only more curiosity), to a true representation, or *sampling*, of nature that might become the proper, privileged place for conducting research into nature" (my emphasis). Nick Wilding, "Graphic Technologies," in *Robert Hooke. Tercentennial Studies*, eds. Michael Cooper and Michael Hunter (Ashgate, Aldershot, 2006), 123–134, 126.

27 Hooke outlined his theory of memory in his *Lectures of Light, in The Posthumous Works of Robert Hooke*, ed. R. Waller (London, 1682/published in 1705).

In a review of recent Bacon scholarship, Brian Vickers proposed to characterise Bacon's work as "an 'encyclopedia' of work needed, of not yet existing knowledge, an encyclopedia of lacunae, as it were, which a new philosophy would fill in."[28] I think the term "encyclopedia of lacunae" could be equally used as a beautifully apt description of Hooke's endeavour. With this notion, we can perhaps understand better the seemingly scattered state of Hooke's writings. To interpret this apparent chaos in the light of the "Spirit of Invention," which was a scientific as well as a poetic method, could therefore help us to understand better the specific bent Hooke gave to the inheritance of Bacon in Restoration natural philosophy.

Hooke's poetics, therefore, deeply connected the questions of invention and of arrangement. But in the process, those categories were deeply redefined. Hooke was offering a very material poetics of the new science: *inventio* was the invention of mechanisms, instruments and experiments; *dispositio* now dealt with the very practical question of how to print and organise the successive findings; *elocutio*, I will now show, was concerned with the question of how to best formulate the precious "matters of fact."

Elocutio

> 'twere much to be wished, that others would take this Method in their Publications, and not torment their Readers with such nauseous Repetitions, and frivolous Apologies, in Method and Volumes do necessitate them to; But would rather inrich the Store-house of Art and Nature with choice and excellent Seed, freed from the Chaff and Dross that do otherwise bury and corrupt it.[29]

Building on the Baconian metaphor of the store-house of nature, Hooke advocated a method of writing which protected the precious findings from equivocations: "Matter of Fact being the Kernel Readers generally desire (at least in these Subjects) it will be so much the readier for use if it be freed from the thick and hard shell of Impertinences."[30]

28 Vickers, Brian, "Francis Bacon and the Progress of Knowledge," in *Journal of the History of Ideas* 53, 3 (1992), 495–518, 504, n. 34.

29 Robert Hooke, "To the Reader," *op. cit.*

30 *Ibid.*

Again, the question of writing is crucial in order to conduct the task of collecting data efficiently. The existing scientific genres, with their redundancies, only impede the process of discovery. This stylistic programme was exemplarily implemented in the text of the *Attempt*. The new accurateness that was reached through reformed instrumentation allowed for the formulation of a "matter of fact," precisely in the way advocated in the preface. In the *Attempt*, the new astronomical fact was condensed in the following sentence:

> 'Tis manifest then by the observations of July the Sixth and Ninth: and that of the One and twentieth of October, that there is a sensible parallax of the Earths Orb to the fixt Star in the head of *Draco*, and consequently a confirmation of the Copernican System against the *Ptolomaick* and *Tichonick*.[31]

The discrepancy between the length of the description of the instrumental procedure and the few lines outlining the result is striking. Here the regime of experimental certainty is based on two successive and opposite stylistic modes: the *copia* of the experimental narrative and the conciseness of the fact. Such stylistic concentration recalls the techniques of the aphorism such as gnomic present tense and assertive style. The aim of such style was to turn the mere hypothesis into a fact. The isomorphism between language and reality was thus to define the new poetics of science, which leads me to the question of the "genre" advocated by Hooke.

The title *Attempt* seemed to announce the modest rhetoric of the experimental essay.[32] It is now clear, however, that Hooke was aiming at an experimental proof. The phrase *An Attempt to Prove the Motion of the Earth by Observations* seems an oxymoron between the tentative style of the Boylian essay and the proof that was aimed at. Here as often, I shall suggest, Hooke was playing with the meaning of words: his "attempt" was less a tentative essay than a new experiment, indeed an assay.[33] In fact, as Simon Schaffer has made

---

31 Hooke, *An Attempt, op. cit.*, 25.

32 According to James Paradis, "Bacon's metaphor, the warehouse, reflects the stylistic object of the new literary enterprise to purge self from text, and the structural object to dismantle the received systems and store what remains in a literature of elements and parts," James Paradis, "Montaigne, Boyle and the Essay of Experience" in *One Culture. Essays on Science and Literature*, ed. George Levine (Madison, University of Wisconsin Press, 1987), 59–91.

33 "In the early seventeenth century, impressed by Michel de Montaigne's *Essais*, Francis Bacon composed a set of similarly titled, briefly tentative moral reflections. Robert Boyle

clear, the word "essay" was then used in the senses of both the literary essay inherited from Montaigne through Bacon, and the experimental essay developed by Boyle. In Hooke's *Attempt*, the experiment at stake is an *experimentum crucis*, once again hinting at a Baconian notion: the *instantia crucis* of *Instauratio Magna*.[34] The displacement from "essay" to "attempt" was akin to the move from Bacon's *instantia crucis* to Hooke's *experimentum crucis*: both signified an essential shift toward a new and determining role given to experiments and instruments. Here again, Hooke radicalises Bacon, or at least interprets his text in a very particular way. The double meaning of the title can now be understood in its full epistemological import. Hooke's observation was less an "essay" than an experimental proof, the "kernel of matter of fact" looked for by any serious philosopher. This is not to say that the notions of *experimentum crucis* and invention as mostly experimental and mechanical were the only correct interpretation of Bacon. Rather, I suggest that it was Hooke's reading of Bacon that yielded this particular trend of Baconianism.

Conclusion

As we know, Hooke was wrong. Later astronomers such as Flamsteed, Cassini and Manfredi contested Hooke's measurements and conclusions in the *Attempt*. When James Bradley replicated Hooke's observations of Gamma Draconis with a better Zenith telescope in 1725–1727, he found that the apparent change of the star's position could not be explained by the annual parallax. Rather, this was an optical effect due to the combination of the finite velocity of light and of the Earth's motion through space that he called the aberration of light.[35]

---

soon adapted this kind of literary technology to his experimental "essays." Since then the English language has somewhat distinguished between essay and assay, between literary work in library or study and the messy labours of workshops, mines or mints. Other languages make the distinction fuzzier, as the paired terms essai/experience and Versuch/Probe indicate," Simon Schaffer, "Public Experiments," in *Making Things Public*, *op. cit.*, 299–300.

34 On the transformation and attribution of the phrase *experimentum crucis*, see Brian Vickers, "Francis Bacon and the Progress of Knowledge" *art. cit.*, 511, n. 45. According to Vickers, the phrase was coined by Robert Boyle, in his *Defence of the Doctrine touching the Spring and Weight of the Air* (1662), referring to Pascal's experiment on the Puy-de-Dôme as "an *experimentum crucis* (to speak with our illustrious Verulam)," in Boyle's *Works*, 6 vols., ed. T. Birch (London, 1772), I, 151.

35 See for instance G. Sarton, "Discovery of the Aberration of Light," *Isis* 16 (1931), 233–239, M. Williams, "James Bradley and the Eighteenth Century 'Gap' in Attempts to Measure

Accordingly, historians have generally treated the *Attempt* as a rather embarrassing text: the failure of a restless natural philosopher. It remains that this tract was a very successful manifesto for a new method and a new poetics of science, showing how to solve the problems of experimental philosophy by the complementary use of texts and tools. My own attempt here has been to thus establish a new perspective by considering together the *Attempt* and the contemporary preface, that is, the combination of a defence of mechanistic proof and of a poetics of invention.

I have tried to show that the principle of the "spirit of invention" lies at the heart of Hooke's endeavour and poetics, and helps to re-evaluate what has generally been considered as Hooke's major fault: his tendency to start things and never to accomplish them. Understood in its spatial and material aspect—the free and serendipitous accumulation of samples—Hooke's project appears as a particularly systematic example of Restoration Baconianism. Following Bacon, Hooke refused the traditional scientific genre of the treatise and experimented with a new mode of writing. The Preface to the *Attempt* offers an articulated poetics that can be summarized in three points: a pragmatic of genre, the association of instrumental and literary technologies, and a poetics of matters of fact as kernels. Ultimately, the inventions, both of nature and art, were to be arranged and displayed in various and beautiful store-houses: printed books, memory, and museums. Hooke was therefore a writer for whom the fabric of the text was an integral part of the beautiful garment he was patiently spinning for science:

> This first Discourse is upon an Observation of Nature, and may therefore be properly referred to that Head, though it contein also somewhat of the Improvement of Art: The second speedily to follow, will more properly be referable to Artificial Improvements, though it will contein also many Observations of Nature; and I design always to make them follow each other by turns, and as 'twere to interweave them, being apart but like the Warp or Woof before contexture, unfit either to Cloth, or adorn the Body of Philosophy.[36]

Annual Stellar Parallax" *Notes and Records of the Royal Society of London* 37 (1982), 83–100 and A. Chapman, "Pure Research and Practical Teaching. The Astronomical Career of James Bradley, 1693–1762," *Notes and Records of the Royal Society of London* 47 (1993), 205–212.

36 Robert Hooke, "To the Reader," *op. cit.*

Appendix

Robert Hooke, An Attempt to Prove the Motion of the Earth by Observations

Made by: Robert Hooke Fellow of the Royal Society.

Senec. *Nat. Qu.* lib. I. cap. 30. *'Nè miremur tam tardè erui quæ tam altè jacent.'*

Reader,

I have formerly in the Preface of my *Micrographia* given the World an account of the founding a Physico-Mechanical Lecture in the Year 1665, by Sir John Cutler, for the promoting the History of Nature and Art. In prosecution thereof, I have collected many Observations both of the one and the other kind, and from time to time (as obliged) I have acquainted the Royal Society at their Publick Meetings, both at Gresham Colledge and Arundel House therewith, by Discourse and Lectures thereupon. Now in order to the further promoting the End and Design of this Lecture, I have complyed, with the desire of several of my Friends (though otherwise not thereunto obliged) to commit divers of those Discourses to the Publick, though of themselves for the most part incompleat, and Essayes or Attempts only upon several Subjects which have no dependencie or coherencie one with another. In the doing thereof, I design to avoid any kind of Method or Order that may require Apologies, Prefaces, or needless Repetitions of what is already known, or might have been said upon that Occasion, or may necessitate me to follow this or that Subject, that doth not some way or other offer itself as it were, and prompt me to the consideration thereof. But because they may possibly admit of some better order hereafter, I design to print them all of the same Volume, that so they mey be, when ranged, either stitched or bound together, and may, as occasion requires, be referred to under the Title of their Number and Page. This way I chuse as the best for promoting the Design of this Lecture; for as there is scarse one Subject of millions that may be pitched upon, but to write an exact and compleat History thereof, would require the whole time and attention of a mans life, and some thousands of Inventions and Observations to accomplish it: So on the other side no man is able to say that he will compleat this or that Inquiry, whatever it be, (The greatest part of Invention being but a lucky hitt of chance, for the most part not in our own power, and like the wind, the Spirit of Invention bloweth where and when it listeth, and we scarce know whence it came, or whether 'tis gone.) 'Twill be much better therefore to imbrace the influences of Providence, and to be diligent in the inquiry of everything we meet with. For we shall quickly find that the number of considerable Observations and Inventions this way collected, will a hundred fold cut-strip those that are found by Design. No man but hath some luckey hitts and useful thoughts on this or that Subject he is conversant about, the regarding and communicating of which, might be a means to other Persons highly to improve them. Whence 'twere much to be wished, that others would take this Method in their Publications, and frivolous Apologies, in Method and Volums do necessitate them to; But would rather inrich the Store-house of Art and Nature with choice and excellent Seed, freed from the Chaff and Dross that do otherwise bury and corrupt it. The communicating such happy Thoughts and Occurrences need not much take up a mans time to fit it for the Press; the Relation being so much the better the plainer it is. And matter of Fact being the Kernel Readers generally desire (at least in these Subjects) it will be so much the readier for use if it be freed from the thick and hard shell of Impertinences. This way also is more grateful both to the Writer and the Reader, who proceed with a fresh stomach upon variety, but would be weary and dull'd if necessitated to dwell too long upon one Subject. There are other conveniencies also in this Method of Communication not less considerable then the former, amongst the rest the securing of Inventions to their first Authors, which 'tis hardly possible to do by any other means; for there are a sort of Persons that make it their business to pump and spy out others Inventions, that they may vend to Traders of that kind, who think they do ingenuously to print them for their own, since they have bought and paid for them. Of this there have lately been some Instances, and more may be expected, if this way prevent not.

When things cannot be well explained by words only (which is frequent in Mathematical and Mechanical Discourses) I adde Schemes and delineatious Descriptions of that kind being easier to be made and understood. As near as I can I omit the repeating things already printed, and indeavour to deliver such as are new and my...being myself best pleased with such usage from other Authors. I have begun with a Discourse composed and read in Gresham Colledge in the Year 1670. when I designed to have printed it, but was diverted by the advice of some Friends to stay the repeating the Observation, rather then publish it upon the Experience of one Year only. But finding that Sickness hath hitherto hindered me from repeating the Tryals, and that some Years Observations have already been lost by the first delay: I do rather hast it out now, though imperfect, then detain it for a better compleating, hoping it may be at least a Hint to others to prosecute and compleat the Observations, which I much long for. This first Discourse is upon an Observation of Nature, and may therefore be properly referred to that Head, though it contein also somewhat of the Improvement of Art: The second speedily to follow, will more properly be referable to Artificial Improvements, though it will contein also many Observations of Nature; and I design always to make them follow each other by turns, and as 'twere to interweave them, being apart but like the Warp or Woof before contexture, unfit either to Cloth, or adorn the Body of Philosophy.

CHAPTER 7

The Looking Glass of Facts

Collecting, Rhetoric and Citing the Self in the Experimental Natural Philosophy of Robert Boyle

Michael Wintroub

> What am I believing in when I believe that men have souls? What am I believing in, when I believe that this substance contains two carbon rings? In both cases there is a picture in the foreground, but the sense lies far in the background; that is, the application of the picture is not easy to survey.
>
> LUDWIG WITTGENSTEIN

Seventeenth-century experimental natural philosophy hinged on a distinction between words and things. The former were considered unreliable and untrustworthy, while the latter, self-authenticating and true. Yet despite a commitment to a thoroughly materialist epistemology, rhetorical practices and discursive conventions were crucial to the conduct of experimental natural philosophy in the early modern period.[1] This rhetoric was not simply an abstract "metadiscourse" outlining the precepts of proper literary style however, but a complex web of social, epistemological, and spiritual practices

1 On the importance of rhetoric for the developing genre of the experimental report in seventeenth-century England, see Steven Shapin, "Pump and Circumstance. Robert Boyle's Literary Technology," in *Social Studies of Science* xiv (1984), 481–520; Steven Shapin and Simon Schaffer, *Leviathan and the Air-Pump. Hobbes, Boyle and the Experimental Life* (Princeton, 1985); Brian Vickers, "The Royal Society and English Prose Style. A Reassessment," in *Rhetoric and the Pursuit of Truth. Language Change in the Seventeenth and Eighteenth Centuries* (Los Angeles, 1985), 3–76; Robert Markley, "Robert Boyle on Language. *Some Considerations Touching the Style of the Holy Scriptures*," in *Studies in Eighteenth-Century Culture*,vol. xiv, ed. O.M. Brack, Jr. (1985), 159–171; Jan Golinski, "Robert Boyle. Scepticism and Authority in Seventeenth Century Chemical Discourse," in *The Figural and the Literal Problems of Language in the History of Science and Philosophy, 1630–1800*, eds. A.E. Benjamin et al. (Manchester, 1987), 58–82; Peter Dear, "Narratives, Anecdotes, and Experiments. Turning Experience into Science in the Seventeenth Century," in *The Literary Structure of Scientific Argument* (Philadelphia, 1991), 135–163; *idem*, "*Totius in verba*. Rhetoric and authority in the early Royal Society," *Isis* lxxvi (1985), 145–161; and John T. Harwood, "Science Writing and Writing Science. Boyle and Rhetorical Theory," in *Robert Boyle Reconsidered*, ed. Michael Hunter (Cambridge, 1994), and his excellent introduction to *The Early Essays and Ethics of Robert Boyle* (Carbondale, 1991), xv–lxix.

 | DOI 10.1163/9789004283701_009

by which the putative distance between words and things could be effaced. This was to be accomplished, as the seventeenth-century French erudite, Pierre Fontanier, explained, by "painting things in such a lively and energetic manner" that they would virtually appear before a reader's eyes, thus "...making of a narrative an image, a picture or even a living scene."[2]

Fontanier's rhetorical strategy of "painting" with words was neither original nor a seventeenth-century invention. It had a long and venerable history. This paper offers a brief history of such rhetorical practices; my aim being to demonstrate the crucial role they played in efforts to articulate the precepts of social and epistemological conduct by which experimental natural philosophers secured the honour, credibility and virtue of their craft and of their knowledge. In particular I will focus on the work of Robert Boyle.

The trope of visual presence—of creating with *words* the impression of actually seeing *things*—featured as a central part of Boyle's efforts to construct convincing accounts of his experimental trials. In utilizing this narrative form he, like Fontanier, exploited a long tradition of rhetorical theory and practice. However, the ability of Boyle's texts to mediate between the literary and the material cannot be attributed solely to the formal discursive conventions he employed, but also to his personal credibility as a gentleman experimenter, a scholar and a Christian. Despite avowals to the contrary, rhetorical practices were crucial in this regard. An important part of what follows, therefore, will be concerned with examining the significance that rhetoric had in articulating and enacting the normative ideals of gentlemanly civil conduct underpinning the social organization of Restoration experimental natural philosophy.[3] My overarching goal, however, will be to untangle the intricately spun web of literary and discursive practices by which words were endowed with the capacity to substitute for the presence of things; and indeed, to examine the ways in which words themselves came to be treated (like the material objects and matters of fact at the centre of natural philosophical discourse) as things to be collected, classified, displayed and deployed, as evidence of an experimentalist's social status, epistemological credibility, and moral trustworthiness.

2 Pierre Fontanier, *Les figures du discours* (Paris, 1968), 390, in Louis Marin, *Portrait of the King* (Minneapolis, 1988), 87.

3 Steven Shapin's *A Social History of Truth. Civility and Science in Seventeenth-Century England* (Chicago and London, 1994) should be consulted in this regard. Also see works cited in ref. 1, especially Shapin's "Pump and Circumstance," and Dear's "Narratives, Anecdotes, and Experiments."

History, Virtue and Truth

In a fragment found in Strabo (*Histories* XXXIV, 4, 4), Polybius, the great historian of the second century BC, points to Homer's vivid and circumstantial prose style as proof of the veracity of his verse:

> Now the end aimed at by history is truth and so we find the poet (Homer) in the Catalogue of Ships mentioning the peculiar features of each place, calling one town "rocky," another "on the border," another "with many doves," another "by the sea"; the end produced by this disposition of details is the 'vividness' of actual presence (*'impression de vie'*), as in his battle scenes, while the myth aims rather to please and astonish.[4]

According to Ginzburg's reading of this fragment,

> in the opposition between history and myth, ...Homer stands firmly on the side of history (and of truth) because the aim (the *telos*) of his poetry is 'vividness' (*enargheian*).[5]

Yet though the *telos* of Homer's verse may have been, as Ginzburg points out, "'vividness' (*enargheian*)," it was also to edify: that is, to inspire the listener to virtuous action.

Speaking generally, ancient historians viewed truth and *virtue* as being synonymous: the truth of history being its ability to persuade the reader to lead an exemplary and virtuous life. To teach by example, to portray in vivid clarity the precepts of right conduct, to call forth the very vision of statesmanship, heroism, and wisdom—this was the purpose of history. This standard of historical truth did not rely on any empirical or referential guideposts, but on the accrediting of a particular style of representation. This was identified by Polybius as *enargeia*—the description of something in such vivid detail that the reader felt that he or she was actually a witness to it.[6] In this sense it was not history, but rather the trope of *enargeia* itself, which Polybius held up as virtuous and worthy of emulation.

Abstracted from the immediacy of its referent (the vivid circumstantiality of Homer's text), *enargeia* became for Polybius a device, a figure, a scheme—a rhetorical skill to be learned and a tool to use. Yet it would be a fundamental

4 As cited in French by Carlo Ginzburg, "Montrer et citer. La vérité de l'histoire," *Le débat*, lvi (1989), 43–54, 45 (my translation).

5 *Ibid.*

6 For example, Perrine Galand, "L'«Enargia» chez politien," *Bibliothèque d'humanisme et renaissance* xlix (1987), 25–53.

mistake to presume from this that *enargeia* was simply a formal characteristic common to certain texts; rather, its use articulated a socially recognized pattern of activity which stipulated the grounds upon which both truth and virtue could be secured. As a mark of distinction, or obversely, of exclusion, *enargeia* was not then merely a stylistic *effet de vérité*, but the defining characteristic of a well-educated and virtuous man. As Quintilian was to put this:

> It is a great gift to be able to set forth the facts on which we are speaking clearly and vividly. For oratory fails of its full effect, and does not assert itself as it should, if its appeal is merely to the hearing, and if the judge merely feels that the facts on which he has to give his decision are being narrated to him, and not displayed in their living truth to the eyes of the mind.[7]

Citing the Evidence

According to Ginzburg's analysis, the rhetorical standard of *enargeia* was supplanted in the seventeenth century by a new empiricist concern for the material evidence of the past. Following Momigliano, he suggests that this shift can be traced to the antiquarian response to sceptical critiques of classical history. Antiquarians reasoned that if the words of ancients could not be trusted, then knowledge of the past could still be discovered through the evidence of coins, medals, buildings, records and statues.[8] No longer the stuff of imaginary heroes, fictional orations, and edifying examples, the field of history became a treasure house of self-authenticating material facts. Thus, Ginzburg explains, standards of truth which relied on *enargeia*, and therefore on rhetoric, were superseded by a stricter, more empirical, emphasis on "facts": *à l'enargeia... s'est substituée l'évidence, la preuve*.[9]

Ginzburg then goes on to suggest that the use of quotation marks is to a culture of print, what *enargeia* was to an oral culture—a means of conveying a specific *effet de vérité*.[10] Both, he argues, shared a similar function, for just as a vivid description recreates in language a picture of the world that seems to

7 Quintilian, *Institutio oratoria*, trans. H.E. Butler (Cambridge, 1966), Book VIII: iii, 61–63.

8 See A. Momigliano, "Ancient History and the Antiquarian," *Journal of the Warburg and Courtauld Institutes* xiii (1950), 285–315.

9 Ginzburg, "Montrer" (ref. 4), 53.

10 *Ibid.* With regard to the history of the footnote from a different perspective see Anthony Grafton's excellent treatment, "The Footnote from De Thou to Ranke," *History and Theory* xxxiii (1994), 53–76.

the reader palpably true, so too quotation marks act to authenticate the evidence displayed in a text. As an example, he cites the sixteenth-century Italian historian, Sperone Speroni, as someone who

> ...recognized that there is a moral obligation associated with quotation marks. They must be unequivocal insofar as they communicate a truth which cannot be altered.[11]

Yet quotation marks and footnotes were also looked upon with some suspicion in both the sixteenth and seventeenth centuries.[12] As the late sixteenth-century erudite, Etienne Pasquier, commented:

> In communicating my present researches to my friends, there were some of them who found it in bad form (*mauvaise grace*) that at the end of every point, I confirmed my words by reference to some ancient author....[13]

Similarly, Francis Bacon argued that one should

> Never cite an author except in a matter of doubtful credit.... And for all that concerns ornaments of speech, similitudes, treasury of eloquence, and such like emptiness, let it be utterly dismissed.[14]

Robert Boyle also shied away from such literary devices as citation. As he put it, "other writers" should not be appealed to as "judges, but as to witnesses," they should not be considered as "oracles" used to confirm the status of one's own opinions, but "as certificates to attest matters of fact."[15] For Boyle citation was not an appeal to the authority of a judge, but to the testimony of a witness. Yet the uneasiness expressed by Bacon, Boyle and the friends of Pasquier indicates

11 Ginzburg, "Montrer" (ref. 4), 52.

12 See Simon Schaffer, "Making Certain," *Social Studies of Science* xiv (1984), 137–152, 140.

13 Cited in G. Huppert, *The Idea of Perfect History. Historical Erudition and Historical Philosophy in Renaissance France* (Urbana, 1970), 34.

14 Cited in James Paradis, "Montaigne, Boyle, and the Essay of Experience," in *One Culture. Essays in Science and Literature*, ed. George Levine (Madison, 1987), 59–91, 69.

15 Boyle's comments are to be found in Thomas Birch (ed.), *The Works of the Honourable Robert Boyle*, 5 vols. (London, 1772), i, 313. Also see Shapin, "Pump and Circumstance" (ref. 1), 496; with regards to the "bad manners" of overly precise authorial citation, see Shapin's *A Social History of Truth* (ref. 3), 117; also see Richard Westfall's "Unpublished Boyle Papers Relating to Scientific Method, II," *Annals of Science* xii (1956), 103–117, especially 113–116.

the precarious status of this distinction, for citation was still strongly associated with the "darkness of schools."

Yet regardless of obvious differences of opinion, there was an overriding similarity of perspective between the critics and the advocates of citation—for both were appealing to an evidentiary standard of proof structured negatively in opposition to rhetorical standards such as *enargeia.* Nevertheless, it would be a great mistake to take the seventeenth-century experimental natural philosopher simply at his word; for despite the explicit rejection of rhetoric, the experimentalist's way of writing, indeed his very identity, was structured by and inscribed within patterns of speech and activity crucially informed by it.[16] And thus, even though he was a vehement critic of rhetorical ornamentation and literary embellishment, Robert Boyle himself admitted that these were a necessary concomitant to his experimental natural philosophy:

> For though a philosopher need not be solicitous, that his style should delight its reader with his Floridness, yet I think he may very well be allowed to take a care, that it disgust not his reader by its flatness, especially when he does not so much deliver experiments or explicate them, as make reflections or discourses upon them: for on such occasions he may be allowed the liberty of recreating his reader and himself, and manifesting, that he declined the ornaments where they may darken as well as adorn the subject they are applied to. Thus…though it were foolish to colour or enamel upon the glasses of telescopes, yet to gild or otherwise embellish the tubes of them, may render them more acceptable to the users, without at all lessening the clearness of the objects to be looked at through them.[17]

The "Weight" and "Matter" of Words

From the fifteenth century such qualities as courtesy and eloquence increasingly became the means by which social and political status could be achieved and maintained.[18] Knowledge of rhetoric was thought to be among

16 On the importance that Boyle placed on rhetoric and eloquence for the "…embellishments of our conceptions and…the congruity of them to our design and method…," see *Some Considerations Touching the Style of the Holy Scriptures*, in *Works* (ref. 15), ii, 301; and also Harwood, "Science Writing and Writing Science" (ref. 1), 44.

17 Boyle, *A Proëmial Essay, Works* (ref. 15), i, 304–305.

18 See Norbert Elias, *The History of Manners. The Civilizing Process*, i (New York, 1978), and also Jacques Revel, "The Uses of Civility," in *A History of Private Life. Passions of*

the defining characteristics of the gentleman. As Thomas Wilson said in his *Arte of Rhetorique*, "Yea, great Lordes would thinke themselves contemned, if learned men sought not to speake in this sort."[19] Commonplace books, manuals and handbooks of courtesy and rhetoric were therefore extremely popular, providing both models and strategies for social advancement and self-understanding.

These works were often called *enchiridions*, which derives from the Greek word for a small hand-held object, such as a dagger.[20] According to Whigham, "as both analyses and weapons, these texts both depict and enact the exercise of power and privilege in a world dominated by the social rhetoric of courtesy."[21] Yet these books were double edged, for in severing the connections between birth and political mastery, and basing élite identity on learnable—and therefore vendible—skills, the door was opened to precisely those people élite groups most wanted to distinguish themselves from. Education was thus both a vehicle for an aspiring class, and the self-definition of a ruling élite which aimed to exclude such parvenus as "...untowardly Assheades, that through malapartness thinke to purchase them the name of a good Courtier."[22] The pejorative connotations of "scheme" (to plot, to intrigue, to make devious plans) become apparent here—i.e., the scheming social climber. As Wilson said, "...hee that can catche an ynke horne terme by the taile, him they coumpt to be a fine Englisheman, and a good Rhetorician."[23]

By way of response to such book-learned gentlemen, the truly noble adopted the cool demeanour of effortless and inimitable nuance, Castiglione's *sprezzatura*, as the mark of their distinction. Display was still required, indeed

the Renaissance,iii trans. Arthur Goldhammer, ed. Roger Chartier (Cambridge, MA, 1989), 167–205. With regard to England see Frank Whigham, *Ambition and Privilege. The Social Tropes of Elizabethan Courtesy Theory* (Berkeley, 1984); Steven Shapin, "Pump and Circumstance" (ref. 1); *idem*, "The House of Experiment in Seventeenth-Century England," *Isis* lxxix (1988), 373–404; *idem*, "'A Scholar and a Gentleman.' The Problematic Identity of the Scientific Practitioner in Early Modern England," *History of Science* xxxix (1991), 279–327; *idem, A Social History of Truth* (ref. 3); Paulo Rossi's *Francis Bacon. From Magic to Science* (London, 1968), 58; Paul Siegal, "English Humanism and the New Tudor Aristocracy," *Journal of the History of Ideas* xiii (1952), 450–468, 466; and D.R. Woolf, Erudition and the Idea of History in Renaissance England," *Renaissance Quarterly* xl (1987), 11–48, 28.

19 Thomas Wilson, *Arte of Rhetorique* (Oxford, 1909), 147.

20 Whigham, *Ambition* (ref. 18), 30.

21 *Ibid.*

22 Wilson, *Arte of Rhetorique* (ref. 19), 162.

23 *Ibid.*

necessitated, but the fact of display was to be hidden.[24] As Thomas Hoby noted in his popular translation of *The Book of the Courtier*, grace requires that one

> ...eschue as much as a man may, and as a sharpe and daungerous rocke, too much curiousnesse, and...to use in everye thing a certaine disgracing [*una certa sprezzatura*] to cover arte withall, and seeme whatsoever he doth and saith, to doe it without painte, as (as it were) not minding it [*quasi senza pensarvi*].[25]

If *sprezzatura* seems a concept alien to the seventeenth-century English gentleman, it is because such explicit, indeed cynical, acts of self-reflexivity were themselves caught up in the general critique of rhetoric and artifice. In this sense, *sprezzatura* was translated into the more congenial terms of plain speaking and humility. The significance of injunctions to speak modestly, however, spiralled beyond gentlemanly acts of anti-rhetorical self-fashioning to dovetail nicely with precepts regulating the proper determination of truth and falsity. Humility, as a species of *sprezzatura*, was thus not only a strategy for creating social distance, but a means by which gentlemen could distance themselves epistemically from those kinds of absolute judgements which were characterized as being vain, scholastic or heretical.

Moreover, if once authoritative truths were now revealed to be false, then those held adamantly in the present might equally prove suspect in the future. Such epistemological relativism was an obvious threat to social, moral and religious order. Epistemic humility enabled the bracketing of such dangers, while identifying those who persisted in unqualified assertions regarding true essences and causes as being at best arrogant, and at worse, heretical. As Obadiah Walker put it, "...if you find a man single in his judgement, be wary of him; he either knows more then all others, or there is some ill principle in him."[26]

The gentleman, on the other hand, was counselled scrupulously to avoid dogmatism and seek rather the continuance of polite civil conversation as the highest of virtues. Thus, for example, in his treatise detailing the specifics of his plan for a college of natural philosophy, Abraham Cowley proposed that the governing body of the college "...lay a considerable pecuniary mulct upon any one who shall be proved to have entered so far into a quarrel as to give

---

24 See, for example, Mario Biagioli's discussion in *Galileo Courtier. The Practice of Science in the Culture of Absolutism* (Chicago and London, 1993), 52.

25 Cited in Whigham, *Ambition* (ref. 18), 93.

26 Obadiah Walker, *Of Education, Especially of Young Gentlemen* (Oxford, 1673), 237, cited in Shapin's *Social History of Truth* (ref. 3), 117.

uncivil Language to his Brother-Professor...."[27] Thomas Sprat mirrored these concerns when in his *History of the Royal Society* he wrote that "...eloquence ought to be banished out of all civil societies, as a thing fatal to peace and good manners...."[28]

If one was to avoid the twin dangers of dogmatism and pride, conversation regarding truth required the thoroughly rhetorical anti-rhetoric of probabilism, humility and modesty.[29] Discourse having to do with absolutes was to be maintained solely with regard to more immediate and self-authenticating—i.e., empirical—paths to knowledge, such as collecting and experimental natural philosophy.[30] As Cowley warned, nothing else could be trusted, for "Our Reasoning Faculty as well as Fancy, does but Dream when it is not guided by sensible Objects."[31]

Seventeenth-century courtesy literature paralleled the natural philosopher's emphasis on *sprezzatura*, humility and material registers of the truth.[32] "[N]othing forced may appeare with that decencie, as when it is naturally descended...," Brathwaite remarked in the *English Gentleman*.[33] Likewise, he

---

27 Abraham Cowley, *The Advancement of Experimental Philosophy* (1661; reprinted in facsimile, Menston, 1969), 41.

28 Thomas Sprat, *History of the Royal Society*, facsimile of the 1667 edition, eds. J.I. Cope and H.W. Jones (Saint Louis, 1958), 111.

29 On the relationship between probabilism and civil conversation see Shapin, *A Social History of Truth* (ref. 3), 117–119, 124–125, and "Pump and Circumstance" (ref. 1), 494–497.

30 See, for example, Michael McKeon, *The Origins of the English Novel, 1600–1740* (Baltimore, 1987), especially chap. 2, "The Evidence of the Senses. Secularization and Epistemological Crisis."

31 Cowley, *op. cit.* (ref. 27), 10.

32 Recently there have been a number of excellent studies which have explored the relationship between the courtly imperatives of *civilité* and the development of early modern experimental natural philosophy and natural history. In addition to works previously cited by Steven Shapin, see especially Mario Biagioli, "Etiquette, Interdependence, and Sociability in Seventeenth-Century Science," in *Critical Inquiry*, vol. xxii (Winter, 1996), 193–238; *idem*, "Scientific Revolution, Social Bricolage, and Etiquette," in *The Scientific Revolution in National Context*, eds. R. Porter and M. Teich (Cambridge, 1992), 11–54; Lorraine Daston, "Baconian Facts, Academic Civility, and the Prehistory of Objectivity," in *Rethinking Objectivity*, ed. Allan Megill (Durham, N.C.,1994), 37–63; Paula Findlen, "Controlling the Experiment. Rhetoric, Court Patronage and the Experimental Method of Francesco Redi," *History of Science* xxxi (1993), 35–64; *idem*, *Possessing Nature. Museums, Collecting, and Scientific Culture in Early Modern Italy* (Berkeley, Los Angeles and London, 1994); Jay Tribby, "Body/Building. Living the Museum Life in Early Modern Europe," *Rhetorica* x (1992), 139–163; and *idem*, "Cooking (with) Clio. Eloquence and Experiment in Seventeenth-Century Florence," *Journal of the History of Ideas* lii (1991), 417–439.

33 Richard Braithwaite, *The English Gentleman* (London, 1630), 83.

commended the gentleman to speak simply, without "gawdie ornaments" or "words of Art (merely invented to delude)." While Henry Peacham, the Younger, wrote in his widely read *The Complete Gentleman*, that

> ...good and eloquent stile of speaking...*is a judicious fitting of choise words, apt and grave sentences unto matter well disposed, the same being uttered with a comely moderation of the voice, countenance and gesture*; Not that same ampullous and Scenicall pompe, with emptie furniture of phrase wherewith the stage, and our pettie Poeticke Pamphlets found so big, which like a net in the water, though it feeleth weightie, yet it yeeldeth nothing; since our speech ought resemble plate, wherein neither the curiousnesse of the Picture, or faire proportion of Letters, but the *weight* is to be regarded...[for] first your hearer coveteth to have his desire satisfied with *matter*, ere hee looketh upon the forme or vinetry of words....[34]

Here Peacham's emphasis on the display of "weight" and "matter," as opposed to the "forme or vinetry of words," directed his readers beyond language to the substantiality of the "evidence" to which a gentleman's words were to refer. Peacham's advice concerning civil conversation thus closely paralleled his admonition that the gentleman occupy his time with such worthy pursuits as collecting antiquities. In this sense, Peacham was entirely sympathetic to both the antiquarian's suspicions concerning the reliability of books as well as to the faith he placed in the historical verisimilitude of material artefacts such as coins and medals. Accordingly, in his *The Complete Gentleman*, he made the collecting of coins and the like into a key component of polite learning:

> ...for bookes and histories and the like are but copyes of Antiquity, bee they never so truly descended unto us: but coynes are the very Antiquities themselves. But would you see a pattern of the *Rogus*, or funerall pile, burnt at the canonization of the Romane Emperors? would you see how the Augurs Hat, and Lituus were made? Would you see true and undoubted modells of their Temples, Altars, Deities, Columns, Gates, Arches, Aquaeducts, Bridges, Sacrifices, Vessels, *Sellae Curules*, Ensigns and Standards, Navall and mural Crownes, Amphytheatres, Circa, Bathes, Chariots, Trophies, Ancilia, and a thousand things more; *Repare to old coynes, and you shall finde them...there you shall see them excellently and lively represented.*[35]

34 Peacham, Henry the Younger, *The Complete Gentleman* (London, 1634), 42–43 (my emphasis is added on "weight" and "matter").

35 *Ibid.*, 123–124.

For Peacham, material artefacts of Antiquity embodied historical truths far more reliably than words ever could. Yet their representational efficacy closely mirrored the rhetorical functions of *enargeia*—that is, as the "excellent and lively representation" of things. In this sense, Peacham translated the credibility of *enargeia* as a rhetorical *effet de vérité* to the ability of material things to speak the truth. Not surprisingly, he also enjoined the gentleman to adopt the methods of the antiquarian—to collect rare, marvellous and curious objects as both signs of social status and as a means of reforming and extending man's useful knowledge of the world.

Precisely these same values were adopted by Peacham's friend, Sir Francis Kynaston, in planning his *Musaeum Minervae*, an academy dedicated to the proper education of the English nobility. Among the studies that Kynaston suggested young gentlemen make was the careful investigation and collection of antiquities, coins and medals.[36] The similarity between Kynaston's and Peacham's proposals regarding proper gentlemanly behaviour can, at least in part, be traced to the close association of both men to Thomas Howard, second Earl of Arundel, and the circle which met at his Strand home, Arundel House.

Gentlemen and Scholars

Lord Arundel was among the most consummate collectors of art and antiquities in seventeenth-century England. As Shapin notes, "the house that contained his collections, was made into a visible symbol of how a cultivated gentleman ought to live."[37] Henry Howard extended his grandfather's field of interests from the cultural to the natural world.[38] After the Great Fire of London, he invited the Fellows of Royal Society to make Arundel House the

36 Francis Kynaston, *The Constitutions of the Musaeum Minervae* (London, 1636), 4–5. Parliament refused to fund this scheme; it was resubmitted several years later by Kynaston's friend and patron, Arundel, but there is no evidence that these plans were ever implemented.

37 Shapin, "The House of Experiment" (ref. 18), 38.

38 This interest, however, seems to have been present as early as the 1620s; as the instrument maker and mathematician William Oughtred says: "I have been still much frequented [while at Arundel House] by Natives and Strangers, for my resolution and instruction in many difficult poynts of Art." See his *Apologie…to the English Gentrie…the Just Apologie of Wil Oughtred…*(1634), B 1 (v°), cited in A.J. Turner, *Of Time and Measurement* (London, 1993), 107.

centre of their activities, even offering part of his estate as a permanent home to their projected college of natural philosophy.[39]

The addition of the Society's Fellows to the already established circles frequenting Arundel House points to the merging of two different sorts of sensibilities, that of the virtuoso who cultivated an attitude of cultural sophistication as an attribute of status and prestige, and that of the scholarly natural philosopher who pursued similar ends through the production of "useful knowledge" of Nature.[40] The collection of art, antiquities, books and manuscripts found in the civil space of Arundel House was exemplary of the curiosity and refinement of the English gentleman. A collection, however, was every bit as important for the scholar as it was for the gentleman. To this end, the cabinet of Robert Hubert, which John Evelyn described as being Europe's "fullest, and certainly noblest collection of natural raritys," was purchased in the spring of 1666.[41] The inclusion of the Society's Repository of natural objects points to the extent to which the worlds of the gentlemanly collector and that of the scholarly natural philosopher had begun to overlap. Indeed, despite Robert Hooke's denigration of collections used for purposes of "Divertisement, ...Wonder, and Gazing" (rather than "for the most serious and diligent study of the most able and Proficient in Natural Philosophy"), the Society's collection of natural artefacts was closely linked to both the cultured antiquarianism of the gentleman and the erudite antiquarianism of the scholar.[42] With regard to the latter, Hooke explicitly appealed to the scepticism motivating the antiquarian's project to legitimate the Society's collection and study of natural objects. As he put it, shells and other natural forms are

---

39 In Thomas Birch (ed.), *The History of the Royal Society of London, for Improving Knowledge from Its First Rise*, ii (London, 1756), for example, 243. Also see in this regard Michael Hunter's "A 'College' for the Royal Society. The Abortive Plan of 1667–8," in his *Establishing the New Science* (Woodbridge, 1989), 156–184.

40 See the classic articles by W.E. Houghton, "The English Virtuoso in the Seventeenth Century," *Journal of the History of Ideas* iii (1942), 51–73, 190–219; also see Shapin's "'A Scholar and a Gentleman'" (ref. 18).

41 Regarding the purchase of this collection see the letter from Henry Oldenburg to Boyle dated 27 January 1666 in *The Correspondence of Henry Oldenburg*, eds. A. Rupert Hall and Marie Boas Hall (Madison and London, 1965–86), iii, 32, cited in Hunter, *op. cit.* (ref. 39), 127. For more details on the Royal Society's collection see Hunter's excellent treatment, "Between cabinet of curiosities and research collection. The history of the Royal Society's 'Repository,'" found in the same collection, 123–155; the citation for Evelyn is on p. 128.

42 Robert Hooke, "A Discourse of Earthquakes," in *The Posthumous Works of Robert Hooke*, ed. Richard Waller (originally published in London, 1705; facsimile with introduction by T.M. Brown, London, 1971), 338.

> the Medals, Urnes, or Monuments of Nature whose Relievoes, Impressions, Characters, Forms, Substances, etc., are much more plain and discoverable to any unbiassed Person, and therefore he has no reason to scruple his assent.... These are the greatest and most lasting Monuments of Antiquity, which, in all probability, will far antidate the very Pyramids, Obelisks, Mummys, Hieroglyphics, and Coins, and will afford more information in Natural History, than those other put altogether will in Civil.[43]

With regard to the former, the natural artefacts in the Royal Society's repository clearly served to both animate and give content to gentlemanly civil discourse.[44] "Wonder and Divertisement" played a crucial role in this function, and remained, despite Hooke's admonitions, a vital part in shaping the Society's collections and the discussions which they inspired. As for example the curiosities donated to the Repository by Sir Robert Harley during the 18 April 1667 meeting of the Society at Arundel House: "1. A young tiger's skull. 2. Four Indian arrows of that kind which is called poisonous. 3. Some strange American flies. 4. An ape's skull. 5. Two fish scales, in part silver-coloured. 6. Two nuts, one of which, bared of its scale, had the likeness of an ape's face. 7. An excellent colour made of the American Rocou-tree, with a leaf of that tree."[45]

The Society's collection, however, was neither simply a spur to curiosity, nor a vehicle for enhancing the reputation and the prestige of its members, but a means of assembling a complete and accurate description of the natural world. In this sense, the Repository constituted the centrepiece of the Society's program of linguistic reform. As a response to anti-rhetorical and sceptical critiques of language, its members sought to use their collection as a means of establishing a natural language which could bridge the gap between the conventionality of words and the brute reality of things; that is, as Sprat put it, to return "back to the primitive purity...when men deliver'd so many things almost in an equal number of words."[46] Accordingly, Robert Hooke explained

43 *Ibid.*, 335, also see 397.

44 Regarding the role that collections played in animating civil conversation in the early modern period see ref. 32 above, and in particular, Findlen's *Possessing Nature*, especially 97–150, and Tribby's "Body/building."

45 Birch, *The History of the Royal Society* (ref. 39), ii, 167.

46 Sprat, *History of the Royal Society* (ref. 28), 113. In this regard, Tony Davies comments that John Wilkins's project aimed at the "abolition of the distinction between words and things." See his "The Ark in Flames. Science, Language and Education in Seventeenth-Century England," in *The Figural and the Literal*, eds. A.E. Benjamin, G.N. Cantor and J.R.R. Christie (Manchester, 1987), 83–102, 87.

that he planned to order the Society's collection "under its several heads, according to the exact Method and Ranks of all the Species of Nature, which had been compos'd by Doctor Wilkins...in his *Universall Language*."[47] Hooke would do this, he said, so that the "Inquirer" might

> ...peruse, and turn over, and spell, and read the Book of Nature, and observe the *Orthography, Etymologia, Syntaxis,* and *Prosodia* of Nature's Grammar, and by which as with a *Dictionary,* he might readily turn to and find the true Figure, Composition, Derivation and Use of the Characters, Words, Phrases and Sentences of Nature written with indelible, and most exact, and most expressive Letters, without which Books it will be very difficult to be thoroughly a *Literatus* in the Language and Sense of Nature.[48]

The Society's collection was a means of solidifying the fragility of a status distinction based upon learnable skills. It did this by linking these skills to the hard and fast materiality of nature over and against the conventionality of the words of men. Yet the Society's collection aimed to remake the gentleman into a *Literatus* not only in the Language of Nature, but in the Language and Sense of Virtue as well.[49] Seen from this perspective, the Society's collection of natural objects was much like Arundel's collection of cultural artefacts, for both sought to articulate the form and the content of civil conversation, and thereby the social and cultural sensibilities which informed the natural philosopher's or collector's identity as a gentleman and as a scholar.[50] This complementarity marked the fulfilment of Peacham the Younger's ideal of vividly displayed gentility. He characterized this as an *enargic* public presentation which combined

47 Sprat, *History of the Royal Society* (ref. 28), 251.

48 Hooke, *The Posthumous Works* (ref. 42), 338.

49 On Arundel and his collection see David Howarth, *Lord Arundel and His Circle* (New Haven and London, 1985), and also Graham Parry, *The Golden Age Restor'd. The Culture of the Stuart Court, 1603–42* (Manchester, 1981), 108–136.

50 On the relationship between practices of collecting and identity formation see ref. 32, and also Susan Stewart, *On Longing. Narratives of the Miniature, the Gigantic, the Souvenir, the Collection* (Baltimore, 1984), 157–158; James Clifford, "Objects and Selves. An Afterword," in *Objects and Others. Essays on Museums and Material Culture* ed. George Stocking Jr (Madison, 1985), 236–246. In this sense, as Pierre Bourdieu suggests, collecting, as a kind of aesthetic consumption of goods, can be compared to other forms of commodity consumption insofar as it serves to distinguish and demarcate an individual's tastes and dispositions as being those of a particular class or group; see, for example, *Distinction. A Social Critique of the Judgement of Taste*, transl. by Richard Nice (Cambridge, Mass., 1984).

both the visible attributes of the learned scholar with those of the cultivated gentleman. As he said:

> Who is nobly borne, and a Schollar withall, deserveth double Honour... *for heerby as an Ensigne of the fairest colours, he is afarre off discerned,* and winneth to himself both love and admiration, heighthning with skill his Image to the life, making it precious, and lasting to posteritie.[51]

Visible Symbols, Literary Precedents and Civil Conversation

Arundel House was indeed a "visible symbol" of how a cultivated gentleman and a scholar ought to live. It was not without precedent however, for it closely resembled the literary model provided by Erasmus's colloquy, the *Convivium religiosum* ("The Godly Feast") written in 1522.

The narrative of Erasmus's text follows the conversations of the Godly feast's host, Eusebius, and his guests as they toured his house and gardens. "Other men have luxurious homes," Eusebius explains, "I have one where there is a great deal of talk...."[52] Nature and art actively intervened and supplemented this "talk." Inscriptions, animals, paintings, plants and texts, were thus vividly displayed as inspirational landmarks animating the feast's discussions of virtue and moral Christian life. Like relics on the itinerary of a pilgrim, these artefacts, both man-made and natural, were collected and deployed in Eusebius's house as a means of motivating and giving content to individual and collective spiritual meditation.

Erasmus does not privilege nature over art in this role. As Eusebius put it: "...we are twice pleased when we see a painted flower competing with a real one. In one we admire the cleverness of Nature, in the other the inventiveness of the painter: in each the goodness of God who gives all these things for our use...."[53] God's voice—whether through natural objects or human artefacts—is literally ever-present in Erasmus's text. As Eusebius tells his companions: "Nature is not silent but talks to us all the time, [teaching]...the observant many things if she finds him attentive and receptive."[54] Yet though Nature's voice resounded throughout the text, it was heard by its interlocutors, as it is for its readers, only through writing; that is, in the authoritative languages of

51 Peacham, *The Complete Gentleman* (ref. 34), 18.

52 Desidarius Erasmus, *Ten Colloquies*, transl. by Craig Thompson (New York, 1986), 135.

53 *Ibid.*, 137.

54 *Ibid.*, 131.

Antiquity in the form of Latin, Greek and Hebrew inscriptions which accompanied the objects, artefacts and specimens displayed at Eusebius's villa.[55]

By juxtaposing natural objects and human artefacts, things and words, Erasmus aimed to elide the commonplace distinction between godly nature and human (and therefore fallible) art. Thus he integrated the spiritual ideal of Man as a contemplative and pious observer, with the ethical ideal of Man as a creative being endowed with the capacity to transform the World. In this sense, the distinction between the nobility (which associated itself with active public life) and the scholar, was mediated by a notion of civically engaged religiosity. Accordingly, at the colloquy's end the archetypal noble pursuit of the hunt is transformed by Erasmus into a quest for the spiritual and political well-being of Eusebius's neighbours and friends. "…I'm hunting something other than boars and stags," Eusebius tells his guests upon taking leave from the "Godly Feast" and its inspirational setting.

> In one village a certain friend lies critically ill. The doctor fears for his body; I fear for his soul, rather. Since, in my opinion, he is scarcely prepared to depart as befits a Christian, I'm going to be at hand with good counsel, so that he may be helped whether he dies or recovers. In another village a serious difference has arisen between two men; not bad men, certainly, but stubborn ones. If the quarrel should become sharper, I fear it would draw many others into a feud. I'll try my best to bring them to good terms again, for I have bonds of long standing with both. These are my quarry. If my hunting goes as I wish, we'll have a victory celebration here without delay.[56]

The text of Erasmus mirrors the house and gardens of his protagonist, for it provides the reader, as Eusebius provides his guests, with worthy examples for spiritual and moral contemplation. It does this by following the colloquy's characters as they interact with each other and with their environment, to build up—through their combined voices—a vivid and circumstantial prose picture of the items collected in Eusebius's house and gardens. Truly, Eusebius's house was, as he claimed it to be, "a feast for both eyes and minds."[57]

The trope of *enargeia* plays an essential role here, for it was used by Erasmus to call forth a living picture of Eusebius's house before his readers' eyes. In this

---

55 *Ibid.*, 133, 137–140.

56 *Ibid.*, 173; on the importance of the metaphor of the hunt for early modern science, see William Eamon, *Science on the Secrets of Nature. Books of Secrets in Medieval and Early Modern Culture* (Princeton, 1994), 269–300.

57 *Ibid.*, 135, 173.

manner, he presents his audience with a palpably real collection of natural and artificial objects capable of inspiring reflection upon God's goodness and the nature of virtuous conduct. Yet his text was not meant simply as an ésoteric exercise in private religious meditation, for just as the "Godly Feast" ended with Eusebius taking leave of his companions to pursue "good works" beyond the sanctified confines of his country retreat, so too, Erasmus must have hoped that his readers would transform their textual reflections upon virtue, truth, and Nature, into useful worldly action.

The ideal of Christian sodality represented by the spiritual discussions and virtuous deeds animated by Eusebius's country villa, exemplified many of the same ideals of virtue and conviviality adopted by the circle of virtuoso gentlemen who met at Arundel's Strand home a century later. Moreover, the gentlemanly conversations which took place at Arundel House were anchored to similar sources—that is, to Arundel's and later, the Society's, collections. Thus Henry Peacham was to describe Arundel House in terms remarkably similar to those employed by Erasmus in describing the country home of Eusebius:

> You shall find all the walles of the house inlayde with them [inscriptions] and speaking Greek and Latin to you. The Garden especially will afford you the pleasure of a world of learned Lectures in this kinde.[58]

With its elaborate gardens, rich collections of art and antiquities, and with the voice of nature and culture ever present in the form of Latin and Greek inscriptions, works of art and natural artefacts, Arundel House embodied the Erasmean ideal of both virtuous civil conversation and the inspirational setting in which it was set. A place in which both nature and art actively informed and taught those who visited it, Thomas Howard's home on the Strand mirrored that belonging to Eusebius in Erasmus's text—it too was a "feast for the eyes and the mind."

Copious Collections and Paper Museums

Erasmus was considered exemplary of both learning and piety by the seventeenth-century English virtuoso.[59] Thus, for example, in a letter written

58 Peacham, *The Complete Gentleman* (ref. 34), 106.

59 See, for example, Shapin's "'A Scholar and a Gentleman'" (ref. 18), 283, and Harwood's "Science Writing and Writing Science" (ref. 1), 45. It is perhaps worth noting that the word

in the summer of 1666, John Beale implored Robert Boyle to follow in Erasmus's footsteps.

> This man [Erasmus] with great judgement gave us the choicest of Grecian and Roman antiquities, for all the ornament and uses of philology, morality, and true theology: restored the best of the Gentiles, as well as Christian fathers, to themselves, recovering them from dust, worms, and worse-defacing scribes, into the light of fairest impressions; and with his smart apologies, smiling reprehensions, and healing lashes, constrained all the nests of lazy monks, as well as the swarms of over-busy sectaries, to attend to a serious reformation, as well of literature and manners, as of faith and notions.[60]

Beale's praise no doubt met with approval by Boyle, who was an ardent admirer of the Rotterdam humanist. Texts by Erasmus figured prominently in the library of Boyle's residence in London, which contained no fewer than ten separate volumes by him, more than by any other single author.[61] These included two copies of the colloquies in which the "Godly Feast" appeared. Another volume in this library was Erasmus's *De duplici copia verborum ac rerum commentarii duo.*

De copia was among the most influential treatises of style and rhetoric in the sixteenth and seventeenth centuries. It elaborated as a formal pedagogical method what the "Godly Feast," which was written some ten years later, came to articulate as a meditative literary exercise. Organized as a collection of schemes

---

virtuoso was first introduced into the English language by Henry Peacham; see *The Complete Gentleman* (ref. 34), 105.

60 Letter from John Beale to Robert Boyle, dated 8 August 1666, Boyle, *Works* (ref. 15), vi, 413.

61 Excepting, of course, works by Boyle, and those by the late seventeenth-century religious reformer, John Owen, of which there were also ten. See Harwood, *The Early Essays* (ref. 1), which lists the contents of what Harwood originally identified as Boyle's library, 252–281. Recent evidence, brought to light by Michael Hunter, has challenged Harwood's identification of this booklist with Boyle's library; rather it seems to have belonged to Boyle's servant and executor John Warr, see Hunter's microfilm edition of the Boyle Papers, *Letters and Papers of Robert Boyle* (1992). Nevertheless, many of the books on this list were kept in rooms at Boyle's home. And indeed, Warr both lived and worked closely with Boyle for some twenty years. Thus, though one cannot directly link this booklist to Boyle's library, it remains a valuable source of evidence for understanding the intellectual milieu within which Boyle lived and worked. I would like to thank an anonymous referee for *History of Science* for calling my attention to Hunter's comments regarding this booklist; I would also like to thank John Harwood for generously sharing his thoughts with me regarding the current status of his original claims about it.

and tropes to be used to dress up, ornament and inspire eloquent speech, its central theme was the correct display of *copia*.[62] The word *copia* referred to abundance, plenty, variety and to the place where this abundance was to be found.[63] *De copia*, the book, was just such a place. So too was the "Godly Feast." Indeed, keeping *De copia* in mind when reading the "Godly Feast," one can see beyond the palpable materiality of the vividly described verbal objects collected at Eusebius's villa, to the trope of *enargeia* itself as being among the principal objects being collected and displayed by the text. Accordingly, like the "Godly Feast," but set out in more explicit terms, *De copia* was a kind of *Kunstprosa*, a paper museum or cabinet whose collected objects consisted of the schemes, tropes, figures and *sententiae* which were to be learned and deployed as evidence of an individual's moral, intellectual and spiritual virtuosity.[64]

In a time characterized by immense social mobility, sectarian violence and political upheaval, evidence of personal credibility and trustworthiness was not founded solely upon the lineage of blood, or on economic wherewithal, but on performative acts of self-citation which referred to a common repertoire of verbal objects collected in such textual cabinets as Erasmus's *De copia* and his "Godly Feast." Rhetorical figures and tropes, such as *enargeia*, were markers authenticating the social, epistemic and spiritual credibility of those who deployed them. As such, they had a social function analogous to that attributed to the footnote by Ginzburg; that is, they were used as a means of conveying a specific *effet de vérité*. Solutions to problems of textual authority (i.e., the footnote), thus found an echo in solutions used to renegotiate the basis of social authority.[65]

Boyle's early works, such as his *Diurnall Observations Thoughts, & Collections*, his *A Diurnall Miscellaneous Collection*, or the *Aretology*, can be firmly located in the early modern tradition of commonplace books to which Erasmus's works belong.[66] As the influential seventeenth-century educator, Comenius, argued, such commonplace books, were to be used "…as general registers of subjects in

62 W.S. Howell, *Logic and Rhetoric in England, 1500–1700* (Princeton, 1956), 117.

63 Terence Cave, *The Cornucopian Text. Problems of Writing in the French Renaissance* (Oxford, 1979), 5–6. Cave notes that "…one of the particular senses of copia is 'treasure-chest,' 'hoard,' or 'store' (thesaurus)."

64 On the *Kunstprosa* in early modern England see, for example, Anthony Grafton, "Barrow as a scholar," in *Before Newton. The Life and Times of Isaac Barrow* ed. Mordechai Feingol (Cambridge, 1990), 291–302, especially pp. 296–297.

65 Desiderius Erasmus, *Collected Works of Erasmus*, ed. Craig R. Thompson, transl. and annotated Betty Knott, vol. xxiv (Toronto, 1978), 295.

66 According to Ann Blair, the commonplace book was "…a crucial tool for storing and retrieving the increasingly unwieldy quantity of textual and personal knowledge that

which we file and arrange all that is worthy of note."[67] Boyle—being well-acquainted with Comenius,[68] as he was with Erasmus—clearly identified with this interpretation, seeing the appropriate display of such textual collections as a clear and direct means of asserting social and spiritual distinction; as he said:

> ...to be able to collect moral and spiritual documents out of a book of hieroglyphicks, or from a landskip or a map, is more than every attentive considerer can do, and is that, which argues something of dexterousness and sagacity, that is not very ordinary.[69]

For Boyle, it was from the evidence of experience, and its meditative translation into writing, that the axiomatic principals defining a gentleman's self-identity were to be founded. As he put this in one of his early essays, *The Dayly Reflection*:

> ...it is consonant to Reason as wel as verify'd by...numerous Presidents; that he who makes it his constant Practise, both to obserue attentively... the considerable Passages...that occurre to him, and not content to treasure them in...his Memory drawes them out...into those Consequences, and mints...them into <those> Axioms they are capable of affording.... For in effect, Experience Consists, not in the multitude of years but in that of Observations.... *Thus this Admirable Examen makes...a man...both the Teacher, the Scholler and the Booke of his owne selfe.*[70]

We see here Boyle's emulation of his most eminent predecessor in the elaboration of English natural philosophy, Francis Bacon. Bacon argued that though

---

guaranteed copiousness in speech and writing." Indeed, she concludes that "[t]he commonplace book thus spread as widely in Renaissance Europe as the Erasmian ideal of eloquence through copia rerum or abundance of material." Ann Blair, "Humanist Methods of Natural Philosophy. The Commonplace Book," *Journal of the History of Ideas* liii (1992), 541–551, 542. Though Boyle's early treatises are not, in the strict sense, commonplace books, they were—as Shapin points out with regard to similar ethical texts—"...fundamentally structured through the rearrangement and resituation of the contexts of personal commonplace books." See Steven Shapin, "Personal Development and Intellectual Biography. The Case of Robert Boyle," *The British Journal for the History of Science* xxvi (1993), 335–345, quotation on 340.

67 Jan Amos Komensky (Comenius), *The Analytical Didactic of Comenius* (1649), trans. by Vladimir Jelinek (Chicago, 1953), 148.

68 Harwood, *op. cit.* (ref. 1), 260.

69 Boyle, *Works* (ref. 15), ii, 340.

70 Boyle, *The Dayly Reflection*, fol. 272 (v°) in Harwood, *op. cit.* (ref. 1), 208, my emphasis.

commonplace books were "…a matter of great use and essence in studying, as that which assureth copie of invention," they were in fact of little worth, for they represented "merely the face of a school, and not of a world…."[71] By way of alternative, he suggested that collections be culled from the book of nature; that is, from experience, instead. Similarly Boyle, riding a wave of anti-rhetorical and anti-scholastic opinion, distanced himself from the textual collections of moral and spiritual *sententiae* which characterized his early works, to base his ideal of moral virtue on the book of nature instead. As he put this in the preface to his *Occasional Reflections*,

> …there is such a difference betwixt him, that but takes up instructions in books of morality and devotion, and him that by occasional reflections derives from the book of nature, and the accidents he chances to take notice of, as there is betwixt an ant, that contributes nothing either to the production or improvement of the corn she lays up and feeds on, but only carries away that, which she finds ready formed into its little granary or repository; and the industrious bee, who, without stealing from flowers anything that can prejudice them, does not only gather, but improve and transform her food, and live on that, which otherwise would be useless….[72]

However, not taking either Bacon or Boyle at their word, the schemes, figures and tropes (*copia verbum*) collected in such books as Erasmus's *De copia*, can nevertheless be directly compared, despite their anti-rhetorical objections, to objects (*copia rerum*) contained in other sorts of early modern collections.

Collections are constituted both in and by the historically specific principles governing their organization and use.[73] They are, in this sense, only meaningful in conjunction with the particular social, cultural and spiritual narratives which organise them. In books of grammar, rhetoric and devotion, such as those by Erasmus, this narrative was contiguous with a social field in which notions of status and legitimacy were rooted in exacting distinctions made on the basis of manners, speech and writing.[74] As Boyle's contemporary, George Snell, said:

71 Francis Bacon, *The Advancement of Learning*, ed. Arthur Johnston (Oxford, 1974), 129–130, cited in Blair, "Humanist methods" (ref. 66), 550.

72 Boyle, *Occasional Reflections* (ref. 69), 340.

73 See, for example, Stewart, *On Longing* (ref. 50), 155–156.

74 The act of showing, of display, as a means of status distinction, implies a parallel transformation in ways of seeing; the theatre of correct manners, dress and speech demands not only that the actor be seen by an audience, but that he too observes—measures, dissects and evaluates—the manners and motives of those around him. See Refs. 18 and 32 above.

> ...persons of a truely gentile, and perfect education, in points of civilitie, can easily perceiv at the first enterview by manner of salutation of obeisance, of posture, gesture and speech, whether a youth have been well educated or not...[75]

It was therefore of the utmost importance that a gentleman, if he be worthy of the title, speak effectively. If he didn't measure up, the consequences could be dire; for he would

> ...bee noted, censured and condemn'd to bee as an unbred Rustick, and senseless Clown; and having once such a disgracing asspersion and report raised of him, hee shall hardly wipe it off for manie years after...[76]

For the nobleman and prince, collecting was a conspicuous and tangible sign of social and cultural superiority, while for those of lesser standing such collections provided a means of access into élite circles.[77] As Lorraine Daston has noted, "a famous collection could draw crowned heads to one's doorstep for a tour, and published catalogues were often frank advertising, sent to various celebrities."[78] And while the correct verbal display of schemes and tropes might not bring crowned heads to one's doorstep, it could serve to advertise the speaker as a man of quality who was eligible to participate in élite circles. Taken in this sense, the words, schemes and tropes found in Erasmus's *De copia* can be compared to the objects in a cabinet of curiosities; they were objectified artefacts—textual "matter"—to be collected, classified, studied and publicly displayed as evidence of a gentleman's standing and credibility.[79] Commonplace books of rhetoric, grammar and courtesy, can thus be understood as being part

75 George Snell, *Right Teaching of Useful Knowledge* (London, 1643), 58–59.

76 *Ibid.*

77 See, for example, Findlen, *Possessing Nature* (ref. 32), 105.

78 Lorraine Daston, "Reviews on Artifact and Experiment," *Isis* lxxix (1988), 452–467, 460. Also see Giuseppe Olmi, "Science—Honour—Metaphor. Italian Cabinets of Curiosities in the Sixteenth and Seventeenth Centuries," in *The Origins of Museums. The Cabinet of Curiosities in Sixteenth- and Seventeenth-Century Europe,* eds. Oliver Impey and Arthur MacGregor (Oxford, 1985), 5–16, and Krzysztof Pomian, *Collectors and Curiosities. Paris and Venice, 1500–1800,* transl. by Elizabeth Wiles-Portier (Cambridge, 1990), especially chap. 1, "Between the Invisible and the Visible. The Collection."

79 See for example, Anthony Grafton and Lisa Jardine, *From Humanism to the Humanities. Education and the Liberal Arts in Fifteenth- and Sixteenth-Century Europe* (London, 1986), 136.

of the same complex of cultural and social practices informing other types of early modern collecting.[80]

Situated within this context, *copia*, as the collection and display of texts, schemes and tropes, partook in the larger field of discursive practices within which the antiquarian, gentleman, and natural philosopher collected, classified, and displayed their medals, coins and matters of fact. By spanning the distance between *res* and *verba*, truth and rhetoric, nature and art, the reification of language as objectified social artefacts to be collected on the page and deployed in civil conversation, exemplified an internal movement within rhetoric itself which mirrored the anti-rhetorical ("empirical") attitudes expressed by antiquarians and experimentalists who sought to escape the deceptions of rhetoric by appealing to the "incontrovertible" weight and substance of the "evidence." In this sense, the seventeenth-century reliance on the evidence of non-literary antiquities and artefacts constituted, as Michael McKeon has pointed out, "…not a flight from the abstraction of script and print but a heightened commitment to it."[81] We can thus perhaps cast *copia*'s semantic net a bit further, from *copia* as a treasure-chest of schemes and tropes, to *copia* as a kind of collection in the cabinet of the self.

The Appropriateness of Truth

Enargeia was one of the key figures found in Erasmus's *De copia*; it formed the centre around which the other schemes and tropes enclosed within the treasure chest of *copia* were to be organized and displayed. According to Erasmus, "…instead of setting out the subject in bare simplicity," the trope of *enargeia* is employed to

> fill in the colours and set it up like a picture to look at, so that we seem to have painted the scene rather than described it, and the reader seems to have seen rather than read.[82]

On the formal level it would be difficult to assess the truth or falsity of such an "*enargic*" account. For Erasmus, however, the relevance of such a determina-

80 Tribby makes a similar point with regards to the homologies existing between "the display of collected objects in a museum and the display of the civil body in early modern Europe …," and the ways in which "a courtier could use a museum to raise his profile and distinguish himself from his contemporaries…." See Tribby, "Body/building" (ref. 32), 152–153.

81 McKeon, *The Origins of the English Novel* (ref. 30), 43.

82 Erasmus, *De copia* (ref. 65), 577.

tion depended crucially on the context in which such a narration was situated. *Enargeia*, he said, is frequently employed in tragedies to present something which "...either cannot be represented on the stage for practical reasons or which is not the sort of thing one wants to represent."[83] In such cases, he concluded, it does not matter whether a narration is true or false, "...as in Sophocles's *Electra*, where the old man gives Clytemnestra a false account of how Orestes lost his life in the chariot race."[84] In other cases, however, he argued that *enargeia* should move beyond the given circumstances of historical chronology to employ more persuasive descriptive language:

> In a description not only do we take in things that happened before, during, and after the incident in question, but *we also point out that something that did not happen could have happened if such and such had come to pass, or could still happen.* One could for example say: See what a narrow escape we all had when you engaged the enemy with so little thought for the consequences: if by some mischance the enemy had been victorious, this and that would have happened. Or someone speaking against the introduction of monarchy could by his description set before the eyes of his hearers the whole drama of tyranny, and urge them to imagine that they see before them what they will soon be actually experiencing if they change from democracy to kingship.[85]

Here Erasmus evokes the situated and engaged character of a narrator trying to persuade his audience over and above a simply denotative assessment of truth by reference to the "evidence." Yet, in his discussion of the proper ways to describe people and ways of life, Erasmus seems to adhere to a more strictly empirical standard of verification. As he put it, "...to express these things well not only requires imagination and skill," but also to "have actually seen with your own eyes what you wish to depict."[86]

Despite these apparent discrepancies in Erasmus's treatment of *enargeia*, there is a common theme running throughout his argument; that is, his repeated emphasis on the absolute importance of eloquence and decorum for civil behaviour. For Erasmus a criterion of appropriateness for the time, place, and circumstances of its use governs *enargeia*'s relationship to the "truth." As he said, "...the fundamental requirement for eloquence is to speak

83 *Ibid.*, 578.
84 *Ibid.*, 579.
85 *Ibid.*, 579, my emphasis.
86 *Ibid.*, 582.

as the occasion demands, and no utterance is well spoken which is lacking in this quality of appropriateness."[87] Not surprisingly, knowledge of the proper use of *copia* was considered the purview of an educated élite. Thus it was that the language of truth and the language of social mastery met on the common ground of correctly displayed *copia*.[88] It was precisely this standard that was employed by Boyle in appraising his own rhetorical practice of *enargeia*:

> I know too, that there may be found, in some of those protases, divers passages, and particularly some descriptions...which to some readers will not seem noble and gaudy enough. *But to such perusers it may be represented, that a suitableness to the theme, how mean soever it be, may very well, as a piece of decorum, be allowed to a writer, and in a few cases more than in point of descriptions: and that these being but pictures drawn (with words instead of colours) for the imagination, the skilful will approve those most, that produce in the mind, not the finest ideas, but the likest;* as a critick in limning will more prize the picture of an old meagre Sibyl, where the wrinkles and the sallow skin are drawn exactly to life, than a dozen ordinary pictures of the spring (which yet are wont to charm vulgar eyes) though the youthful face, which represents that florid season, have as gaudy colours upon the cheeks and lips, as imbellish the roses and lilies, which compose the chaplet that adorns the head.[89]

The Rhetoric of Evidence

De copia went through 85 editions during Erasmus's lifetime, and many more there-after. There were a number of English editions, including the first, in 1512, which Erasmus dedicated to his close friend John Colet, dean of St Paul's School in London. The importance of this book cannot be overestimated. Virtually every work of pedagogy written in sixteenth-century England advocated its reading,[90] and most works on rhetoric and courtesy liberally

87 *Ibid.*, 559.

88 Indeed, for Erasmus, such intellectual virtue was definitive of nobility. As he said: "... everyone who cultivates the mind in liberal studies must be taken to be noble. Let others paint lions, eagles, bulls, or leopards on their escutcheons; those who can display 'devices' of the intellect commensurate with their grasp of liberal arts have a truer nobility." See his *De civilitate. De civilitate morum puerilium*, in *The Erasmus Reader* ed. Erika Rummel (Toronto, 1990), 102.

89 Boyle, *Occasional Reflections* (ref. 69), 329, my emphasis.

90 See, for example, Peacham's *The Complete Gentleman* (ref. 34), 28.

borrowed from it.[91] Two such works were Richard Sherry's *Treatise of Schemes and Tropes*, and Henry Peacham the Elder's *Garden of Eloquence*. Both these works include discussions of *enargeia*.

Sherry defined *enargeia* as "evidence or perspicuitie called also description rhetoricall…when a thynge is so described that it semeth to the reader or hearer…he beholdeth it…."[92] Sherry's equation of *enargeia* to evidence can be traced to Quintilian's *Institutio oratoria* where *enargeia* was translated into Latin as *evidentia* or *evidentia in narratione:* "…when a truth requires not merely to be told, but to some extent obtruded…."[93] The distinction between *res* and *verba*, truth and rhetoric, which is characteristic of the modern commonplace usage of the word 'evidence,' clearly separates us from Sherry, who saw 'evidence' as being synonymous with 'description rhetoricall.' For Sherry, as with Erasmus, *enargeia*, correctly displayed, was the unmistakable sign (the "evidence") of one who possessed the requisite status and virtue to speak truthfully. Indeed, Sherry identified effective speech not as an accurate portrayal of external reality, but as the repudiation of the idioms of everyday life.[94] As he explicitly put this, a scheme, such as *enargeia*, "…is the fashion of a word, sayynge, or sentence, otherwyse wryten or spoken then after the vulgar and comen usage…."[95]

Peacham the Elder's *The Garden of Eloquence* appears to follow Sherry closely here; "Schemates Rhetorical," he says,

> be those figures or forms of speaking, which do take away the wearisomnesse of our common speech, and do fashion a pleasant, sharpe, and evident kind of expressing of our meaning; which by the artificiall forme both give unto matters great strength, perspicuitie and grace….[96]

Though Peacham's encyclopaedic compilation of schemes and tropes is undeniably similar to Sherry's,[97] there are hints that within his book "evidence" was beginning to split apart into the "artificiall forme" of rhetoric on the one hand,

91 Fritz Caspari, *Humanism and the Social Order in Tudor England* (Chicago, 1954), 33.

92 Richard Sherry, *A Treatise of Schemes and Tropes* (Gainesville, 1961), 66.

93 Quintilian, *Institutio* (ref. 7), Book IV: ii, 64. Also see Ginzburg, "Montrer" (ref. 4), 46.

94 Howell, *Logic* (ref. 62), 127–128. Also see, for example, Friedrich Nietzsche, *The Genealogy of Morals*, transl. Francis Golffing (New York, 1956), 160, regarding what Nietzsche called "the pathos of nobility and distance."

95 Sherry, *Treatise* (ref. 92), 25.

96 Henry Peacham, *The Garden of Eloquence* (Gainesville, 1954), 40.

97 Peacham's analysis borrows extensively from Sherry's *Treatise*, to the extent that he directly copies certain passages; see Howell, *Logic* (ref. 62), 135.

and into the "matters" (the empirical evidence) upon which rhetoric was to act, on the other. Though *enargeia* does not appear by name, Peacham the Elder's figure of *descriptio* is strikingly similar to it. In fact, these terms were quite closely related, for the Greek word *ekphrasis* from which *descriptio* was translated, had as its proper end, its *telos, enargeia*.[98] As he said:

> Descriptio…is when the Orator by a diligent gathering together of circumstances, and by a fit and naturall application of them, both expresse and set forth a thing so plainly and lively, that it seemeth rather painted in tables, then declared with words, and the mind of the hearer ther by so drawen to an earnest and stedfast contemplation of the thing described, that he rather thinketh he seeth it than heareth it.[99]

The idea of evidence, in the modern sense, seems to be emerging in Peacham's account as something separate from its rhetorically vivid display—that is, from *descriptio*; for on the one hand we have the "circumstances," and on the other, the "fit and natural application of them." Peacham here seems to be moving beyond mere stylistics to a concern for the "…aptness and truth of the circumstances," to the distanciation of the evidence as a separate entity to which rhetoric is applied as would be an ornament. Yet insofar as the "circumstances" are subordinated to the "scheme" with which the orator presents himself, its "truth" remains tied to propriety and "aptness" of speech. Here, as with Erasmus's *De copia*, the empirical referent of the evidence is subordinated to its appropriate display.

Witnessing the Evidence

Though he vehemently criticized the "scenical pompe" of his father's schemes and tropes, Henry Peacham the Younger nevertheless made one such scheme, *efficacy*, the defining quality of effective speech. However, we should not let this change in terminology side-track our analysis of *enargeia*. The connection between *enargeia* and *efficacy* is a strong one; it can be traced to *enargeia*'s similarity to the Aristotelian term, *energeia* (act, activity, energy).[100] These two words were often treated as variant spellings of one another; perceived as being identical, their respective meanings in fact merged. Thus, for example,

98 See Ginzburg, "Montrer" (ref. 4), 49.

99 Peacham, *The Garden* (ref. 96), 134.

100 Ginzburg, "Montrer" (ref. 4), 45.

Richard Sherry, when he discusses *enargeia* in his *Treatise of Schemes and Tropes* used both spellings on the very same page.[101] Similarly, as Terence Cave notes, the different words *enargie*, *energie* and *eficace* were used, respectively, by Du Bellay, Sebillet and Peletier in a manner which leaves little room to doubt that they were taken to mean the same thing, namely *enargeia*.[102] Thus, though Phrissemius's *scholia* on Agricola's *De inventione* III.3 distinguished between *enargeia*, translated as *evidentia*, and *energeia*, rendered as *efficacia*,[103] Henry Peacham the Younger did not make such a distinction, for his definition of *efficacy* unmistakably points to *enargeia*. As he says,

> *Efficacie*...is a power of speech, which representeth a thing after an excellent manner, neither by bare words onely, but *by presenting to our minds the lively Ideas or formes of things so truly, as if wee saw them with our eyes*....[104]

Despite the various changes in terminology there remained a recognizable core of meanings traceable through these various incarnations of *enargeia* (i.e., *energia*, *efficacia*, *efficacy*, *enargeia*, *evidentia*, *evidence*). As mentioned earlier, Quintilian translated *enargeia* into Latin as *evidentia*. Not withstanding the spatial and temporal distance from its forbear, the English word "evidence" clearly resembles its ancestor. Accordingly, the first entry in the *O.E.D.* under "evidence" is from Robert Boyle's *Occasional reflections*: "Certain truths that have in them so much native light or evidence...it cannot be hidden." Though *enargeia* had all but disappeared by the time Boyle wrote these words, his definition of evidence as "native light" is immediately recognizable in *enargeia*'s primary root, *enarges*, which means "clear" or "palpable," and in the closely related adjective *arg*, which means "bright" or "shining." Argos, after all, had one hundred eyes, some of which were always open.[105]

Thus, though it did not appear by name in Boyle's writings, *enargeia* was maintained under a kind of anti-rhetorical "erasure." In this sense, it was not just a word, or a rhetorical device, but a complex web of social, cognitive and spiritual practices by which the difference between words and the actual presence of things was to be effaced. Accordingly, *enargeia* was also a technique of making knowledge mobile; as such it was a means of extending the validity of

101 Sherry, *Treatise* (ref. 92), 66.

102 Cave, *Cornucopian* (ref. 63), 28.

103 *Ibid.*

104 Peacham, *The Complete Gentleman* (ref. 34), 84 (my emphasis).

105 Ginzburg, "Montrer" (ref. 4), see 46, also see Cave, *Cornucopian* (ref. 63), 28.

truths and practices beyond the specific localities of their production through the multiplication of the experience of being an actual witness.

When Peacham the Younger used the plural possessive ("…by presenting to our minds the lively Ideas or formes of things so truly, as if wee saw them with our eyes"), it was because truths—the evidence—needed to be caught in the parallaxed gaze of many witnesses. Utilizing the "literary technology" which Steven Shapin has identified as "virtual witnessing," Boyle sought to produce in his readers' minds "such an image of an experimental scene as obviates the necessity for either its direct witness or its replication."[106] Accordingly, Boyle's experimental reports aimed at setting forth an exacting circumstantial picture of his experimental trials. Such prose pictures were used to conjure up a vision in the reader's mind of the actual *presence* of an experiment on the written page. To this end Boyle declared the necessity of being "somewhat prolix"—to deliver things so as "…to make them more clear, in such a multitude of words, that I now seem even to myself to have in diverse places been guilty of verbosity…."[107] He felt that despite its stylistic drawbacks, this circumstantiality was necessary to the production of convincing accounts, especially with regard to those readers who lacked the wherewithal to conduct experiments themselves. Boyle's goal in crafting his experimental reports was the multiplication of the number of possible witnesses to his experiments by producing in writing the visual impression of witnessing them. As he put this:

> …the most ordinary reason of my prolixity was, that foreseeing, that such a trouble as I met with in making those trials carefully, and the great expense of time that they necessarily require…will probably keep most men from trying again these experiments, I thought I might do the generality of my readers no unacceptable piece of service, by so punctually relating what I carefully observed, that they may look upon these narratives as standing records in our new pneumatics, and *need not reiterate themselves an experiment to have as distinct idea of it, as may suffice them to ground their reflexions and speculations upon.*[108]

106 Shapin, "Pump and Circumstance" (ref. 1), 491. John Harwood associates the idea of virtual witnessing with Hooke's strategies for verbal and visual *enargeia*, see his "Rhetoric and Graphics in *Micrographia*," in *Robert Hooke. New Studies*, eds. M. Hunter and S. Schaffer (Woodbridge, 1989), 119–147, especially pp. 135–147.

107 See Shapin, "Pump and Circumstance" (ref. 1), 492–493, and Boyle's *A Proëmial Essay* (ref. 15), 302.

108 Boyle, *Works* (ref. 15), i, 2; also see Boyle's *A Proëmial Essay* (ref. 15), 305, and Shapin, "Pump and Circumstance" (ref. 1), 493.

Witnesses, for Peacham, as for Robert Boyle, were not isolated monads, but took part in a very specific form of life—a social grouping of self-fashioned gentlemen who shared certain common values and practices. Thus Boyle announced in the "Advertisement to the Reader" prefacing his *Motives and Incentives to the Love of God* (more generally known by the title *Seraphic Love*) that insofar as his work was written to a young gentleman who was "…a traveller, a linguist, and a scholar, it was not improper to discourse with him at the rate of supposing him so qualified."[109] The figure of *enargeia* was one measure of this qualification. Its use was wrapped-up in the self-presentation and understanding of the proper gentleman; as such, it was an immediately recognizable sign of membership and status within this community. Boyle's goal, following Erasmus, Sherry and Peacham, was to convey through his text an illusion of verisimilitude, to craft a textual account of an experimental trial which would present "…to our minds the lively Ideas or formes of things so truly, as if wee say them with our eyes…." Such a vivid representation signified to the reader that an account was both credible and trustworthy.[110] When Boyle set out to write credible and trustworthy accounts of his experimental trials and the matters of fact which they produced, he thus appealed to common and recognized standards of respectability associated with gentlemanly behaviour.

As a social and epistemological category, *enargeia* was both the identifying mark of a gentleman and the guarantee of his credibility—as an experimentalist and as a producer of matters of fact. As one courtesy writer put it, "the promise and word of a Gentleman is construed of the like validitie that his deede, confirmed with the contestation of witnesses and seale, could extend unto…."[111] If an experimental report was written in the appropriate manner (that is, *enargically*) it could be taken on trust that the things it recounted had actually happened. In writing up his experimental reports, Boyle was thus appealing to recognized standards of effective speech, to *enargeia*, in order to extend the immediacy of the witnessing experience—and his own authority—beyond the walls of his laboratory. The credibility of these accounts was ensured insofar as they were carefully crafted to appeal to—while at the same time articulating—the specific idioms which characterized gentlemanly civil conversation. Accordingly, Boyle's experimental reports extended beyond mere reportage of matters of fact, to actively intervene to create, legislate and

109 Boyle, *Works* (ref. 15), i, 246.

110 See Peacham, *The Complete Gentleman* (ref. 34), 24, and Brathwaite, *The English Gentleman* (ref. 33), 83–84.

111 Cited in Whigham, *Ambition* (ref. 18), 72.

shape the discursive conventions and social practices definitive of the natural philosopher in Restoration England. Yet, like Erasmus before him, he also understood rhetoric—and specifically, *enargeia*—in spiritual and moral terms.

The Virtuoso's Glass

In his dialogue "Upon the sight of a looking-glass with a rich frame," Boyle argues, under the name of Eugenius, that the rich frame surrounding a looking glass can be compared to the eloquence used by a sermon to attract its auditors. Without eloquence, says Boyle, the sermon is like a looking-glass with but a "mean or common frame": it is left "unregarded, though it be for substance excellent…."[112] He goes on to explain that though a frame "…be not part of the glass, nor shews us any part of our face," it does,

> by its curious workmanship, attract our eyes, and so invites us to consult the glass, that is held forth in it; so the wit, and fine language, wherein [a sermon] is dressed up, though it be no essential or theological part of the sermon, yet it is often that which invites men to hear, or read it.[113]

The thesis of Eugenius is then taken up and refined by the character of Lindamor, who adds that though

> a curious frame doth as well please, as attract, the eye without representing to it the lively image of the beholder's face; so the fine expressions you applaud, are commonly parts of a sermon, that have no specular virtue in them; I mean that have no power, like a good looking-glass, to acquaint the beholder with the true image or representation of his own complexion, and features; nor will this gaudy frame shew him what is otherwise than it should be; the discovery of which, nevertheless, in order to the rectifying what is amiss, is the principle and genuine use of a looking glass. And therefore, as no skilful man will judge of the goodness of a glass, by the fineness of the frame, but rather by its giving him a true representation of his face, without liking it the worse, for showing him its moles, and warts, or other blemishes, if it have any; so no wise Christian will judge of a rousing sermon, rather by the language, than the divinity,

112 Boyle, *Occasional Reflections* (ref. 69), 386.
113 *Ibid.*, 386.

> or will think the worse of a good book, for discovering his faults, or making him think the worse of his own, or other men's ill courses.[114]

Speaking through the character of Eusebius (which, it should be recalled, was the name of the protagonist of Erasmus's the "Godly Feast"), Boyle concludes his meditation on the looking glass by condemning those who would confuse a glass with its frame, or a sermon with its rhetorical embellishments. Accordingly, he argues that stylistic artifice posed the danger of undermining the efficacy of a sermon by diverting it from its primary purpose—to inspire its auditors to reflect upon God's goodness and upon their own moral and spiritual well-being. Eusebius thus condemns those priests who, out of their concern for "winning favour," would rather appear as "an orator than a divine."[115] While for those "witty and critical auditors" who would rather "judge" a sermon than "judge themselves by them," he holds out only contempt. Such men, he says, are like those who would concern themselves with "the embossed images of the frame, without caring to make use of the glass itself..." to better themselves.[116]

However, in his *Some Considerations Touching the Style of the Holy Scriptures*, Boyle takes a very different view of rhetoric, for here he acknowledges its (*enargic*) power to make present to the reader's mind the transcendent reality of God's goodness. In this sense, Boyle understands rhetorical style not simply as an ornamental embellishment to be likened to an elaborate frame of a looking glass, but rather, to the glass itself. Thus he sees style neither as a useless accessory or a decorative vanity, but "...in that larger sense, wherein the word style comprehends not only the phraseology, the tropes and figures made use of by a writer, but his method, his lofty or humble character...and in a word, *almost the whole manner of an author's expressing himself*."[117]

The Two Edg'd Weapon

Homer was blind. When his writing evoked the clarity of visual presence (*enargeia*), it was because he was directly transcribing the words of his Muse. The divine source mediating the presentation of reality in Homer's verse finds

114 *Ibid.*, 386.

115 *Ibid.*, 387.

116 *Ibid.*

117 Boyle, *Some Considerations* (ref. 16), 256; also see Markley, "Robert Boyle on Language" (ref. 1), 162.

an early modern analogue in Erasmus's linkage of *enargeia*'s efficacy as a bridge between *res* and *verba* to the presence of Christ in his narrative. As Cave notes, the itinerary of Erasmus's dialogue is a "paradigm of actively deployed *enargeia*."[118] The use of *enargeia* in Erasmus's text, Cave continues, is "the catalyst for a series of reflections on the location of a true presence."[119] The regulative principle propelling the itinerary of "The Godly Feast" and actuating its discussion of virtue was the vivid presence of Christ—as evidenced by the lively representation of him in the paintings adorning Eusebius's chapel and library, the holy inscriptions decorating his copious garden, and even in the meal itself, which was unambiguously structured as a sacramental celebration of Christ's transubstantiation. The festive incarnation of the God-head in Erasmus's dialogue thus forms the link between the verbal surface of the dialogue and its intercalation with the sensuous reality of nature and art as displayed energically in the text.

Similarly, for Boyle, who claimed to have "scarce looked upon any other book than that of nature" (and, of course, the *Scriptures*) because of the "weakness" of his eyes,[120] it was the ineluctable presence of God as Nature which underwrote the power of his vivid language to communicate the social, moral and spiritual truth evidenced by his experimentally produced matters of fact. In this sense, the literary technology of *enargeia* was understood by Boyle to be both a means of calling forth God's real presence—i.e., as nature, and as a method by which to induce the reader to reflect upon his moral and spiritual identity as an experimentalist, a gentleman, and a Christian.

Yet such rhetoric also posed a very real danger, for it was, as Boyle said, a "two edg'd weapon...as well applicable to the service of Falsehood as of Truth."[121] Indeed, Homer's Muse could be a devil as well as a god. For Boyle, the pious understanding which rhetoric was capable of eliciting was jeopardized by the possibility that it might also contribute to its opposite—to raving. Raving can be characterized as the predisposition of the active mind to take off on flights of spurious fancy.[122] This potential could be triggered by *enargeia*; or as John Harwood has put it, "...by *visual images evoked through reading, scenery, or memory*..."[123]

118 Terence Cave, "*Enargeia*. Erasmus and the Rhetoric of Presence in the Sixteenth Century," *L'Esprit créateur* xvi (1976), 5–19, 10.

119 *Ibid.*, 16.

120 Boyle, *A Proëmial Essay* (ref. 15), 317.

121 Boyle, *Works* (ref. 15), i, 126; also see Markley, "Robert Boyle on Language" (ref. 1), 167.

122 For example, Harwood, *The Early Essays* (ref. 1), 195.

123 *Ibid.*, xlviii, my emphasis.

Boyle was greatly concerned by this possibility. He considered raving, and the kind of knowledge founded upon it (as for example, that of the sectary), a dangerous threat to the stability of social, moral and spiritual order. He sought to escape the dangers implicit in *enargeia* by espousing an anti-rhetorical materialist epistemology which purported to concern itself only with matters of fact. He did this, of course, rhetorically. That is, by presenting his readers with the vivid (*enargic*) impression of actually seeing sensible objects on the written page.[124] Thus it was that by repeated reflection upon morally appropriate and civically useful objects (such as those produced by his experimental trials), that Boyle hoped to evade the possibility of raving.[125] As he put it:

> ...the safest way to hinder our Thoughts from overflowing to those objects they shud not approach, is to dig them Channels many and deepe enuf for them all to flow in. I meane to entertaine our Thoughts with som thing or other that has a Mixture both of profit and Deliht, as well to satisfy as benefit them. I cannot but wonder men shud take so much Pleasure to spend their whole Thoughts vpon vnprofitable Objects, when there is so great a variety of those that do no less content then they do Instruct.[126]

In writing accounts of his experiments and the matters of fact they produced, Boyle deployed the literary technology of "virtual witnessing," here identified with the ancient trope of *enargeia,* as a means of creating in writing not simply witnesses, but the material *presence* of morally suitable things upon which the Christian virtuoso was to reflect for his—and his community's—moral and spiritual well-being.

124 As Shapin has pointed out, despite the supposedly public character of experimental natural philosophy, in practice, accessibility to experimental trials was extremely limited. Thus, insofar as he sought to affirm the credibility of experimentally produced matters of fact by appeal to a public court of like-minded gentlemanly natural philosophers, Boyle had no choice but to rely upon what he considered a fundamentally flawed medium—that of language. See "Pump and Circumstance" (ref. 1).

125 This strategy, of course, had a long tradition in devotional tracts dedicated to image-assisted meditation. As David Freedberg notes, "[b]y concentrating on physical images, the natural inclination of the mind to wander is kept in check, and we ascend with increasing intensity to the spiritual and emotional essence of that which is represented in material form before our eyes...." See his *The Power of Images. Studies in the History and Theory of Response* (Chicago and London, 1989), 162. With regard to Boyle's use of similitude and metaphor as aids to seeing, remembering and meditating upon the physical world for spiritual ends, see Harwood's "Science Writing and Writing Science" (ref. 1), 87.

126 Harwood, *The Early Essays* (ref. 1), 196; the parenthetical additions to Boyle's manuscript (included in Harwood's text) have been omitted here.

By utilizing this *enargic* method, Boyle positioned himself as a medium of both a divine and a social message: mainly, that men could be moved by spiritual knowledge of nature towards a knowledge of themselves as being both civically engaged moral beings and gentlemanly natural philosophers.[127] Like Quintilian and Erasmus before him, he did not distinguish between knowledge and virtue, but viewed understanding and good works to be essentially and thoroughly conjoined civic duties. He thus saw natural philosophy both epistemically as a means of enriching "the understanding with variety of choice and acceptable truths," and as an instrument which would enable the "willing mind to exercise goodness beneficial to others."[128] In this sense, rhetoric, and in particular *enargeia*, was constitutive not only of the notion of evidence as it developed in the early modern period, but played an integral role in arbitrating and defining the social, epistemological and moral fields within which the "experimental form of life" was produced and maintained.

---

127 Idleness and solitude were among the principle causes of the "Waking Dreame[s]" and "Sinfull Thoughts" characteristic of raving. The rigorous pursuit of a civil calling was, Boyle advocated, "a souveraigne Preservative agenst Idleness (that mother of Vices) and an excellent preuention [against] <of> a world of Idle, Melancholick and exorbitant thoughts, and unwarrantable actions." *Ibid.*, 85.

128 Boyle, *Works* (ref. 15), ii, 5.

CHAPTER 8

Science versus Rhetoric?

Sprat's History of the Royal Society *Reconsidered*[*]

Tina Skouen

In November 1660 a group of English natural philosophers decided to form a society for the promotion and advancement of experimental knowledge. They soon took the name of the Royal Society and hired a young author to write a book about their project. Thomas Sprat's *History of the Royal Society of London* appeared in 1667, and by the end of the century, the Society in London had established itself as Europe's leading centre of science, alongside the Académie Royale des Sciences in Paris (established in 1666).[1] Among the founding members of the Royal Society were the linguist and theologian John Wilkins (1614–1672), the astronomer and architect Sir Christopher Wren (1632–1723), and the chemist Robert Boyle (1627–1691). Whereas John Wilkins is most known today for his attempt to create a new, universal language, Christopher Wren was the one who designed the new St. Paul's Cathedral following the Great Fire of London in 1666. As regards Robert Boyle, he is best known for having developed a gas law that carries his name ("Boyle's law"). Another prominent figure in the early Royal Society, the mathematician and linguist John Wallis (1616–1703), is recognized as a pioneer in the development of a new English grammar that was based on the vernacular instead of Latin.[2]

Although there had been meetings in the circles of both John Wilkins (at Oxford) and John Wallis (in London) since the time of the Civil War in the 1640s, Thomas Sprat's *History of the Royal Society* gives an impression that

---

* An earlier version of this article, written in Norwegian, appeared in *Rhetorica Scandinavica* 47 (2008), 9–29. I want to thank the referees of both *Rhetorica* and *Rhetorica Scandinavica*.

1 See P. Dear, "The Meanings of Experience," in *Early Modern Science*, eds. K. Park and L. Daston (Cambridge: Cambridge University Press, 2006), 108–131 (p. 130). The most comprehensive overview of the events leading up to the establishment of the Royal Society is that provided in H. Aarsleff's entry on "Thomas Sprat," in *Dictionary of Scientific Biography*, vol. 12, ed. C.C. Gillespie (New York: Scribner, 1975), 580–587. This article also offers a summary of the contents of Sprat's *History of the Royal Society*. See also the facsimile edition of *History of the Royal Society* (1667), vols. xii–xxxii ed., with an Introduction, by J.I. Cope and H.W. Jones (Saint Louis: Washington University Studies, 1958), 53–60, Appendix A.

2 L.C. Mitchell, *Grammar Wars. Language as Cultural Battlefield in 17th- and 18th-Century England*, vol. 6 (Aldershot: Ashgate, 2001), 17–32 (esp. p. 19).

 | DOI 10.1163/9789004283701_010

the growth of the new science was directly linked with the restoration of the monarchy in 1660. "For the Royal Society had its beginning in the wonderful pacifick year, …when our Country was freed from confusion, and slavery," writes the later Bishop Thomas Sprat, with reference to the period of civil wars (beginning in 1642), followed by republican rule (1649–1653) and the rule of Oliver Cromwell as Lord Protector (from 1653 to 1658).[3] The establishment of the Royal Society in 1660 thus signals a new beginning, both of knowledge and of the Kingdom. Did it also mark the beginning of a new period in the history of rhetoric in England?

In his book about the Royal Society, Thomas Sprat (who was himself a member) expresses his concern with what he considers to be an excessive emphasis on eloquence. It would be better to ban all use of figurative expressions and to have the mathematical system of signification serve as an ideal. These claims, which we shall presently look at in some detail, have made Sprat's *History of the Royal Society* into the most frequently quoted work as concerns the relationship between rhetoric and science in seventeenth-century England. Does Thomas Sprat make a clear break with the Ciceronian tradition? And did the rise of the new science have any direct consequences on the ancient art of rhetoric? Now that the Royal Society has been celebrating its 350th anniversary, it is time to raise these questions again, not only to review the previous discussions about the views on language that are expressed in *The History of the Royal Society*, but also to reconsider the basis of Sprat's praise of science and critique of eloquence. My argument is that Sprat's *History* represents a continuation of—not a total rejection of—the ancient rhetorical tradition.

A Twofold Strategy

Before closing in on what Sprat says about eloquence, I would like to give an overview of the contents of his work, as well as of the means by which he seeks support of the Royal Society. *The History of the Royal Society of London, for the Improving of Natural Knowledge* (which is the work's full title) actually contains little information about the Society's beginnings. As Thomas Sprat admits

3 Quotation taken from *The History of the Royal Society* (hereafter referred to as *The History*), cited in n. 1 above, p. 58. Due to the inconsistent use of italics in the facsimile edition, I have removed all italics except in cases where they clearly serve to underline the sense. As regards the use of initial capitals, I generally retain them only in quotations that are set as separate paragraphs.

in his prefatory "Advertisement to the Reader," the title is slightly misleading, since he is not just concerned with the history of the Royal Society but also with the history of science.[4] The 438-page-long book is divided into three, the first part reflecting Sprat's ambition to produce an account of "the state of the Ancient Philosophy" as well as of the subsequent development of scientific studies (pp. 1–51).[5] The second deals more specifically with the undertakings of the Royal Society (pp. 52–319). Much space is given to documenting the kind of data that had been collected and the great number of experiments that had been conducted. Part two also contains a number of reports that various members had written on a variety of subjects, including astronomical and anatomical observations, expeditions to foreign countries, and methods on how to make gunpowder, how to dye clothes and grow oysters. In the third and final part of the book (pp. 321–438), Sprat argues more generally in favor of "experimental knowledge" and "new sciences."[6] This is his most difficult task, for as Sprat points out in the very beginning of his work, "there is still much prejudice remaining on many [men's] minds, towards any [new] Discoveries in Natural Things."[7]

How is Sprat going to convince his contemporaries that the Royal Society stands for something new without having anyone suspect it represents a threat to the establishment? Sprat employs a twofold strategy, focusing partly on how the new science may be of benefit to modern society and partly on long-established ideas about the origins of Western civilization. The latter most clearly proves Sprat's mastery of rhetoric, but he also shows great skill with his utilitarian argument, for which he could find support in the Royal Charters that were granted to the Royal Society in 1662–1663.

4 Sprat also makes an excuse that there are some inconsistencies in his work, the reason being that "much of this discourse was written and printed above two years before the rest." For details, see the Introduction by Cope and Jones in *The History*, xiii–xvi.

5 Quoted from Sprat's "Advertisement to the Reader."

6 Quotations taken from the "Advertisement to the Reader" and *The History*, 86. Sprat alternatively uses the term "experimental philosophy," as in *The History*, 25–26. As regards the word "science," Sprat also uses this word in connection with his defence of a "practical philosophy" and of a "new" as opposed to "the old philosophy," in *The History*, 437–438. On p. 118, he also talks about "the sciences of [men's] brains." In the seventeenth-century, "science" (Lat. *scientia*) was not a term denoting studies of nature (i.e. natural sciences) but it could rather be used to designate "any body of properly constituted knowledge," as S. Shapin observes in *The Scientific Revolution* (Chicago: University of Chicago Press, 1996), 5 n. 6. The term "scientist" had not yet been invented.

7 *The History*, 4. For examples of the kind of negative reactions that the Society was facing, see *The History*, Appendix B.

In the first Royal Charter of 1662, Charles II had expressed his particular interest in the kind of experimental studies that could contribute either to striking out on new paths or to improving upon the old philosophy.[8] At the same time, Charles asserted that it had long been his intention "to promote the welfare of arts and sciences, as well as that of our territories and dominions," in the hope that he would be regarded as being "not only the defender of the faith, but the patron and encourager of all sorts of useful knowledge."[9] In the second Charter dating from 1663, Charles even declares himself as being "the founder and patron" of the Royal Society, proclaiming that when they join forces to improve "the knowledge of natural things, and useful arts by experiments," this will not only be of great benefit to the English, but it will be "[t]o the glory of God, and the good of mankind."[10]

Taking his cue from the Royal Charters, Sprat describes Charles—who was actually responsible for an aggressive extension of colonial power—as a philanthropic liberator: "For, to increase the Powers of all Mankind, and to free them from the bondage of Errors, is Greater Glory than to enlarge Empire, or to put Chains on the necks of Conquer'd Nations."[11] Just as the King has managed to liberate his country from the dark spell that was cast on it in the period from the Civil War to the Restoration (described as "twenty years [of] melancholy"), the scientists are seeking to free the human mind from all those obscurities that have been clouding the understanding at least since the middle ages.[12]

---

8 Cf. *particulari autem gratia indulgemus philosophicis studiis, praesertim iis quae solidis experimentis conantur aut novam extundere philosophiam, aut expolire veterem*. There is a link to the original Latin text as well as a full English translation at http://royalsociety.org/Charters-of-the-Royal-Society (accessed Nov. 27, 2009).

9 Quoted from *The History*, 134, where Sprat takes some liberty in translating the Latin charter, which does not refer specifically to any "useful knowledge," but only to "useful arts by experiments" (*utilium scientias experimentorum*).

10 *The History*, 134. This quotation shows that Sprat's English translation is not only based on the first Royal Charter of 1662, as the editors Cope and Jones assume in their note on *The History*, 133 (see *The History*, Notes, 14). For neither the King's declaration that he is the Society's founder and patron (*Fundatorem et Patronum*) nor his assertion that the endeavor will be "To the glory of God, and the good of mankind" (*in Dei Creatoris gloriam et generis humani commodum*) appear in the first charter, but they rather stem from the second charter dating from 1663. As Sprat acknowledges in *The History*, 133, his translation is best considered as an "epitome" of the two "letters patent" that have so far been issued.

11 Quoted from the second page of the unpaginated "Epistle Dedicatory" in *The History*.

12 Inserted quotation taken from *The History*, 58. The Scholastics are described in derogative terms in *The History*, 15–22.

Their common goal is no less than to "redeem the minds of men, from obscurity, uncertainty, and bondage."[13]

Significantly, the process of enlightenment and liberation that Sprat is outlining does not involve a total rejection of the existing learning.[14] Referring to how the seeds of the Royal Society were first planted at Oxford—"that most venerable seat of antient learning," where John Wilkins and his group held fort throughout the Civil War—Sprat assures his readers that "the same men have now no intention, of sweeping away all the honor of antiquity in this their new design."[15] All they ask is that they be allowed to put the existing knowledge to the test through observation and experimentation.[16]

In order to explain how the Royal Society relates to the past, Sprat turns to metaphor: when we are standing by our ancestors' graves, we must pay them due respect, but this does not mean that we should shut ourselves inside their tombs, as if we, too, were dead.[17] In another equally striking metaphor, Sprat argues that there are two ways in which one can preserve a dead man's name: either by having a portrait made that bears his exact resemblance, or by observing his children, which are not exactly like him, but which give new life to the lineage.[18] As Sprat implies, most people would probably agree that making babies is a better method of reproduction than making portraits.[19] In the same way, Sprat reasons, "[i]t is best for the philosophers of this age to imitate the Antients as their children: to have their blood deriv'd down to them; but to add a new Complexion, and Life of their own."[20]

Later in the book Sprat describes the progress of science in allegorical terms as an ongoing voyage of discovery: the map does not always correspond with the territory, and it can also happen that the voyager is misled by false assumptions, as when Columbus set sails for the clouds thinking he was heading for the mainland.[21] Nevertheless, he kept going until he found the truth, and the

---

13 *The History*, 58. Cf. 152.

14 See esp. Section xxii ("A defence of the Royal Society, in respect of the Antients") towards the end of part one, *The History*, 46–51, and also Section i–iv in the opening of part three, *The History*, 321–329.

15 *The History*, 54.

16 See esp. Section xviii ("Their conjecturing on the Causes"), *The History*, 100–109, and cf. 28–29, 50.

17 *The History*, 48.

18 *The History*, 51.

19 *The History*, 51.

20 *The History*, 51.

21 *The History*, 108–109.

same drive also motivates the Royal Society, Sprat says.[22] Further extending his analogy between scientific discovery and expeditions at sea, Sprat maintains that it is highly important that one is able to put the material collected into good use as soon as one returns to shore, or else it will be as if the whole voyage had ended in shipwreck and the cargo was left to corrode with rust.[23] In order to prevent such "miscarriages" from happening, the Royal Society has endeavored to continuously develop new methods and technologies, while at the same time its members build on accumulated knowledge.[24] Thus, when the society's members aspire to categorize every aspect of nature or creation, they envision a world that is hierarchically ordered according to the Aristotelian principles of "the inanimate, the sensitive, the rational, the natural, [and] the artificial."[25] The new scientists' ambition thus emerges as firmly rooted in the old tradition.

Sprat also makes a point that the Society is heavily indebted to the Renaissance humanists for having interpreted the Greco-Roman text corpus. This was "a work of great use," especially since it means that Sprat's contemporaries will not have to spend all their time poring over books but can "prosecute new inventions" instead.[26] Nevertheless, it is important to know what is in the books, because he who remains ignorant of the past, remains a child, Sprat says—with reference to Cicero.[27] The ancient texts can still play an important role, provided that the right use is made of them: whereas the humanists aimed to collect and assemble whatever material they could get from the ashes of the dead, the new scientists will rather spread the ashes about, so that the earth can yield new fruit, Sprat argues.[28] He also emphasizes that every time

---

22 *The History*, 108–110.

23 *The History*, 109.

24 *The History*, 109.

25 *The History*, 110. See also 61. Rather than building on any specific work by Aristotle, Sprat presents an unorganized list of categories known from Aristotelian natural philosophy and physics. While the vegetative plane is missing from Sprat's overview, he has added the categories of "the artificial" and "the natural," most likely to account for Aristotle's division between nature and artifact in *Physics* 2.1.

26 *The History*, 24. Cf. *The History*, 436, where Sprat dismisses "our plodding everlastingly on the ancient writings."

27 I.e. "Tully," *The History*, 25. Sprat's modern editors, Cope and Jones refer to *Orator* 34.120 (*The History*, Notes, 6). Sprat cites Cicero in Latin in *The History*, p. 44 (*Brutus* 64.228); *The History*, 333 (*De oratore* III. 23.88); *The History*, 423 (*Brutus* 24—a reference that is missing from the Cope and Jones edition, because they were not able to identify it, see *The History*, Notes, 64: "The passage has not been found in Cicero").

28 *The History*, 24–25. See also *The History*, 436.

and age should be free to decide what kind of knowledge should be passed on from one culture or generation to the next and what is better left behind.[29] As Sprat points out, Pythagoras, Plato and Aristotle did not start from scratch either, but they took what they wanted from the Egyptians.[30] Why should not we be allowed to behave in the same way as the Greek philosophers? Sprat asks rhetorically:

> Did not they trust themselves, and their own Reasons? Did not they busie themselves in inquiry, make new Arts, establish new [tenets], overthrow the old, and order all things as they pleas'd, without any servile Regard to their Predecessors?… And why shall not we be allow'd the same liberty, to distinguish, and choose, what we will follow?[31]

In Sprat's view, it is perfectly reasonable that the natural philosophers of the Royal Society should treat Aristotle in the same way as he treated his predecessors. It is entirely up to the Society to decide which parts of the classical legacy they would like to keep and which to reject.

What, then, can be said about Sprat's own position as a writer? Did he handle the existing tradition in his own field of language and rhetoric with the same mixture of disrespect and respect as the natural philosophers had done with regard to their predecessors? The dominant view has been that Sprat makes a clear breach with the tradition: according to the standard history of English rhetoric by Wilbur S. Howell (1956, 1971), Sprat's *History of the Royal Society* marks the end of Ciceronian rhetoric.[32] As we shall see, today's scholars are less absolute in their conclusions. The aim of the present study is to point out how we may even treat Sprat as a "rescuer" of rhetoric. Although he does give voice to linguistic views that appear incompatible with the Ciceronian tradition, his argument also represents a continuation of the very same tradition, in so far as it is based on ideas that are fundamental to Ciceronian rhetoric. Adapting the natural scientists' approach to previous learning, Sprat puts the rhetorical tradition to the test, both criticizing and experimenting with it for the sake of making new discoveries.

---

29 *The History*, 25.

30 *The History*, 49. See also 5–6.

31 *The History*, 49–50.

32 W.S. Howell, *Logic and Rhetoric in England, 1500–1700* (Princeton: Princeton University Press, 1956), 388–390; *Eighteenth-Century British Logic and Rhetoric* (Princeton: Princeton University Press, 1971), 482–487.

The New Hercules

Sprat may safely be described as a classically trained writer. He had his M.A. from Oxford, the same "venerable seat of antient learning" that had housed the Royal Society's founding member, John Wilkins, who was the warden of Wadham college—the college Sprat attended.[33] Sprat had thus spent most of his school days gaining mastery of the Latin language and of the rhetorical discipline, and in *The History of the Royal Society* he sourly describes this form of education as a year-long practice in linguistic trick-making.[34] At the same time he would hardly have been commissioned to write *The History of the Royal Society* if he had not been a highly skilled rhetorician. Complying with convention, Sprat proceeds to his task with humility, appealing to his readers' good will by professing that he is neither the first nor the best when it comes to arguing in favor of scientific studies. "[M]y weak hands, ...[are] inforc'd by the eloquence of those excellent men, who have gone before me in this argument," Sprat says, further insisting that no one can compete with Francis Bacon (1561–1626), with respect to either eloquence or learning.[35] "He was a man of strong, cleer, and powerful imaginations," Sprat writes, adding that Bacon's style was also very striking, his imagery having a powerful effect without being bombastic.[36] Even though Bacon's comparisons were sometimes outlandish, they were never so far-fetched as to appear obscure ("[t]he comparisons fetch'd out of the way, and yet the most easie").[37]

Bacon served as the Royal Society's lodestar, and Sprat accordingly describes him in highly appreciative terms. The pioneers of the empirical and experimental sciences are the greatest heroes of our age, Sprat argues, "for they must devest themselves of many vain conceptions, and overcome a thousand false images, which lye like monsters in their way."[38] Just like Hercules, the

33 Quotation taken from *The History*, 54. Wilkins was also the one who recommended Sprat for membership of the Royal Society. For details, see *Dictionary of Scientific Biography*, vol. 12, cited in n. 1 above, 580.

34 *The History*, 112.

35 *The History*, 4–5, 35–36. The editors and Jones point out that Sprat was also heavily indebted to Robert Boyle's *Sceptical Chymist* (1661), see *The History*, Notes, 4. See also *The History*, 421, where Sprat says he hopes his own arguments in favor of experimental science will inspire "the many eloquent and judicious authors, with whom our nation is now more abundantly furnish'd than ever."

36 *The History*, 36. We may note that Bacon figures alongside the portraits of Demosthenes, Cicero, and Sir Philip Sidney (1554–1586) on the frontispiece to Thomas Blount's *The Academy of Eloquence* (1654).

37 *The History*, 36.

38 *The History*, 36.

mythological super-hero, Bacon was faced with gigantic beasts and monsters.[39] There is an explicit reference to Hercules in the first book of Bacon's own work, *The Advancement of Learning* (1605), where the Greco-Roman hero emerges as a model of restraint when it comes to appreciating outer embellishment: when Hercules saw an extravagantly made picture of the beautiful Adonis in a temple, his reaction was one of protest and not praise.[40] As Bacon points out, if there is too much ornamentation, this can prevent our deeper understanding of the matter at hand, and therefore, Bacon says, "there is none of Hercules' followers in learning...but will despise...affectations."[41]

Whereas the Renaissance rhetoricians had hailed Hercules as being a master with words and a perfect example that eloquence can be used in the service of goodness, Bacon and his followers adjusted this image by claiming that Hercules' power lay in his ability to see through words.[42] According to Sprat, the Royal Society, too, is primarily seeking to have "a bare knowledge of things," and "to separate the knowledge of nature, from the colours of rhetorick, the devices of fancy, or the delightful deceit of fables."[43]

This aim is also emphasized in the ode "To the Royal Society" that serves as a preface to Sprat's *History*.[44] In this ode, which was written by the poet Abraham Cowley (1618–1667), both Francis Bacon and the Royal Society are said to have performed a task equal to that of Hercules.[45] The author of *The History of the Royal Society* is also warmly recommended. Judging from Cowley's ode, Sprat has done the art of writing as much service as the Society's natural philosophers have done the art of thinking:

> As [they] from all Old Errors free
> And purge the Body of Philosophy;
> So from all Modern Folies He
> Has vindicated Eloquence and Wit.[46]

---

39 *The History*, 35. See also 29, where Hercules is mentioned by name in a fairly similar context.

40 *The Works of Francis Bacon* [1857–1874], vol. 3, eds. J. Spedding, R.L. Ellis, and D. Denon (facsimile edition, Stuttgart: Friedrich Frommann Verlag, 1963) 284–285.

41 *The Works of Francis Bacon*, vol. 3, 284–285.

42 For the special significance of the Hercules figure in the Renaissance, see W.A. Rebhorn, *The Emperor of Men's Minds. Literature and the Renaissance Discourse of Rhetoric* (Ithaca: Cornell University Press, 1995), 217–218.

43 Quotations taken from *The History*, 40, 62.

44 The ode "To the Royal Society" is preceded by Sprat's dedicatory epistle "To the King" and followed by Sprat's "Advertisement to the Reader."

45 See "To the Royal Society," IX, ll. 1–4.

46 "To the Royal Society," IX, ll. 7–10.

In the lines that follow, it becomes clear that the poet Abraham Cowley is mainly referring to Sprat's prose style, but Sprat also deserves some praise for the shrewd way in which he handles such praise in his text. Most likely alluding to Cowley's prefatory ode, Sprat says that he hopes his readers will forgive him if his prose does not turn out to be as good as expected.[47] For, since the Royal Society have not taken any interest in "fine speaking" or in "the artifice of words," it only seems fair that Sprat should not have made it his first priority either.[48] As Sprat innocently explains: whereas in France, they have long since decided to found an academy for language and literature for the sake of, as it were, polishing their French, the English have rather been concerned with finding out more about the natural world.[49] Put simply, while France has prioritized "the advancement of the elegance of speech," England has focused on "the advancement of real knowledge."[50] Being proud that England was able to establish an academy of science before France, Sprat makes it appear as if the French are lagging behind. When Sprat compliments his French colleague, Paul Péllisson-Fontanier, on having written such an "elegant" book about the Académie française that nothing can compare with it, this compliment too, is laced with irony.[51] If anyone should think that Sprat's *History of the Royal Society* is not as well written as Péllisson-Fontanier's *Relation contenant l'histoire de l'Académie françoise* (1653), the reason may be that people in England generally have more important things to do than to spend hours at their desk, thinking about how to best express themselves. Later in the book Sprat says outright that the English are more interested in hearing what others think about the truth in what they say, than in being complimented on their style.[52] Like the new Hercules, they despise affectations.

Sprat does mention that the Royal Society had been discussing whether or not one should establish the same kind of linguistic academy in England as in France, but he reckons this plan was all too vague for it to deserve much coverage in his book.[53] In fact, the Society never got any further than to organize a linguistic committee that does not seem to have had much influence.[54]

47 *The History*, 40.

48 Quotations taken from *The History*, 112, 40.

49 Whereas the Académie française was first established in 1635, the Académie des sciences was founded in 1666, six years later than the Royal Society. Cf. *The History*, 56.

50 *The History*, 40, 2. Compare pp. 125–126.

51 *The History*, 40.

52 *The History*, 114.

53 *The History*, 44.

54 Sprat was appointed as a member of this committee, along with, among others, the writer John Dryden (1631–1700) and the diarist John Evelyn (1620–1706). For more details,

Why, then, has the early Royal Society often been associated with certain normative views on language? Part of the explanation is that Thomas Sprat offers a very strong recommendation in his *History of the Royal Society* as to the kind of style and diction that should and should not be used.

The Battle of Language

The few passages in *The History of the Royal Society* that deal specifically with "Their manner of discourse," clearly indicate that Sprat and his fellow members of the Society are generally skeptical of what they see as an excessive emphasis on rhetorical skills in their own society.[55] Sprat not only points out that eloquence is a weapon that can be extremely dangerous if it falls into the wrong hands, but he also claims that the incessant word-flow in most fields of learning threatens to undermine the more serious initiatives of the Royal Society.[56] If the Society's members had not been very careful in their own choice of words, their entire undertaking would have fallen apart:

> Unless they had been very watchful to keep [the manner of their discourse] in due temper, the whole spirit and vigour of their Design, had soon been eaten out, by the luxury and redundance of speech. The ill effects of this superfluity of talking, have already overwhelm'd most other Arts and Professions; insomuch, that when I consider the means of happy living, and the causes of their corruption, I can hardly forbear…concluding, that eloquence ought to be banish'd out of all civil Societies, as a thing fatal to Peace and good Manners. To this opinion I should wholly incline; if I did not find, that it is a Weapon, which may be as easily procur'd by bad men, as good: and that, if these should onely cast it away, and those retain it; the naked Innocence of vertue, would be upon all occasions expos'd to the armed Malice of the wicked.[57]

see O.F. Emerson, "John Dryden and a British Academy," in *Essential Articles for the Study of John Dryden*, ed. H.T. Swedenberg Jr. (London: Frank Cass, 1966), 263–280. The prospect of having a formal institution that could oversee the language was first raised in England in the 1570s, according to G. Knowles, *A Cultural History of the English Language* (London: Arnold, 1997), 111.

55 Part two, section xx, *The History*, 111–115.

56 *The History*, 111.

57 *The History*, 111.

Sprat is not only convinced that the obsession with rhetoric can have as many negative effects on the new kinds of scientific study as it has had on most other arts and professions, but he is even tempted to suggest that all use of rhetorical language should be banned. The only thing that keeps him from advocating such a measure is his fear that the good and the bad would have responded very differently, the good obligingly laying down their weapons and the bad refusing to do the same, so that the society would have been rendered less and not more safe than before.[58]

Yet Sprat's indignation is not directed against the rhetorical discipline as such, nor even against the use of tropes and figures in general. As Brian Vickers has emphasized, Sprat is mainly concerned with the excessive use of rhetoric, and his loud warnings against its potentially destructive power are politically motivated.[59] More recently, Ryan J. Stark has argued that Sprat primarily seeks to distinguish between "plain" and "bewitching" tropes and figures: put simply, he scorns the magic power of words, thinking that words should be used merely to illustrate and decorate the substance of an idea.[60] Sprat is anti-occult, not anti-rhetoric.[61]

He has no doubt that the tropes and figures were once used in the best interest of both society and philosophy, but in his own day and age, this is no longer the case:

> [T]he Ornaments of speaking...were at first, no doubt, an admirable Instrument in the hands of Wise Men: when they were onely employ'd to describe Goodness, Honesty, Obedience; in larger, fairer, and more moving Images: to represent Truth, cloth'd with Bodies; and to bring Knowledg back again to our very senses, from whence it was at first deriv'd to our understandings. But now they are generally chang'd to worse uses: They make the Fancy disgust the best things, if they came sound, and unadorn'd: they are in open defiance against Reason; professing, not to hold much

58 *The History*, 111.

59 B. Vickers, "The Royal Society and English Prose Style. A Reassessment," in *Rhetoric and the Pursuit of Truth. Language Change in the Seventeenth and Eighteenth Centuries*, eds. B. Vickers and N.S. Struever (Los Angeles: William Andrews Clark Memorial Library, 1986), 3–76 (esp. pp. 6–7).

60 R.J. Stark, *Rhetoric, Science & Magic in Seventeenth-Century England* (Washington, DC: The Catholic University of America Press, 2009), 51–52.

61 Vickers, "The Royal Society and English Prose Style. A Reassessment," in *Rhetoric and the Pursuit of Truth. Language Change in the Seventeenth and Eighteenth Centuries*, eds. B. Vickers and N.S. Struever, cited in n. 59 above, 52–57; *Rhetoric, Science & Magic in Seventeenth-Century England*, cited in n. 60 above, 53.

> correspondence with that; but with its Slaves, the Passions: they give the mind a motion too changeable, and bewitching, to consist with right practice.[62]

When the poets and philosophers of the antiquity used figurative expressions, they only did so in order to illustrate what they found to be true, or else for the sake of making human knowledge more readily accessible, by appealing to those very senses from which it had derived in the first place.[63] But Sprat thinks his contemporaries are using such imagery for much worse purposes: because the figures no longer appeal to reason but only to the emotions, they can make the most serious thought or subject appear low and cheap. Sprat also plays on Stoic-Ciceronian conceptions about how vehement passions can create so much disturbance in the soul that it becomes quite impossible to think clearly.[64]

Sprat seems genuinely outraged by the thought of the damage that has been caused by such improper use of language in the fields of philosophy and scientific study:

> Who can behold, without indignation, how many mists and uncertainties, these specious Tropes and Figures have brought on our Knowledge? How many rewards, which are due to more profitable, and difficult Arts, have been still snatch'd away by the easie vanity of fine speaking? For now I am warm'd with this just Anger, I cannot with-hold myself, from betraying the shallowness of all these seeming Mysteries; upon which, we Writers, and Speakers, look so bigg. And, in a few words, I dare say; that of all the Studies of men, nothing may be sooner obtain'd, than this vicious abundance of Phrase, this trick of Metaphors, this volubility of Tongue, which makes so great a noise in the World. But I spend words in vain; for the evil is now so inveterate, that it is hard to know whom to blame, or where to begin to reform. We all value one another so much, upon this beautiful deceipt; and labour so long after it, in the years of our education: that we cannot but ever after think kinder of it, than it deserves.[65]

---

62 *The History*, 111–112.

63 According to Cope and Jones (*The History*, Notes, 12), the same argument was voiced by another member of the Royal Society, Joseph Glanvill (1636–1680), in *Plus ultra, or, The Progress and Advancement of Knowledge Since the Days of Aristotle* (1668). Cf. *The History*, 6: "the first masters of knowledge…were as well poets, as philosophers"—a commonplace claim that is especially linked with Sir Philip Sidney's *Defence of Poesie* (1583).

64 Cicero, *Tusculan Disputations*, 3.6.13, 3.7.15, 5.15.43.

65 *The History*, 112.

In Sprat's opinion, everyone who can read has been brainwashed into believing that eloquence is the most important thing of all. This is highly unreasonable, he argues, considering the fact that nothing comes more easily than to learn those tricks that make writers and public speakers seem so great. Hoping to make his readers see that there is no reason why they should continue to admire such ancient tricks, Sprat uses the oldest trick in the world—he makes it appear as if he is virtually shaking with fury ("For now I am warm'd with this just anger, I cannot with-hold myself…").[66] Whereas writers such as Sprat ("we writers") are allowed to give free scope to their emotions in order to move and persuade their audience, everyone else should exercise moderation, especially if they are involved in politics or scientific studies.

To take politics first: at the beginning of his work Sprat announces that eloquence tends to thrive in a commonwealth or republic, where everything is decided by the use of the voice.[67] In Parliament, everyone will be shouting at the top of their lungs, and the power generally belongs to whoever is most capable of swaying the crowd. To Sprat, both *copia* and *controversia* are associated with political instability, and when he talks about "this vicious abundance of phrase, this trick of metaphors, this volubility of tongue, which makes so great a noise in the world," he is clearly alluding to the Puritan leaders, whose powerful way with words he suspects was a major cause of the uproar that led to the Civil War.[68]

Whereas all forms of political and theological debate are potentially risky, Sprat considers scientific discourse as entirely safe, because when one keeps to discussing nature rather than "civil business" and "humane affairs," it is quite possible to disagree without becoming enemies.[69] Scientific debates tend to be intellectual and "unpassionate" rather than emotionally disturbing, at least as long as both sides stick to the truth and refrain from using such rhetorical tricks as they have learnt at school.[70] In general, Sprat considers it best not to employ any rhetorical devices at all, unless, that is, one aspires to become a

66 *The History*, 112. The subsequent quotation is from the same page.

67 *The History*, 19.

68 Quotation taken from *The History*, 112. Cf. Sprat's reference to "the inchantments of Enthusiasm" in his account of the Civil War in *The History*, 54, and to the "the fury of Enthusiasm," in *The History*, 428. For the association of rhetoric with religious fanaticism, see T.M. Conley, *Rhetoric in the European Tradition* (New York: Longman, 1990), 168; S. Irlam, *Elations. The Poetics of Enthusiasm in Eighteenth-Century Britain* (Stanford, California: Stanford University Press, 1999), especially 23–24, 36, 40–45, 51.

69 *The History*, 55–56 (p. 56). See also 70, 429.

70 *The History*, 53–57 (p. 55). See also 82.

great writer: "in most other parts of learning, I look on it [i.e. the use of rhetoric] to be a thing almost utterly desperate in its cure," he says.[71] He then provides a list of the various means by which the Royal Society has attempted to repair the damage that has been done to their field of learning, which brings us to the most controversial feature of Sprat's *History*, namely its proposed reform of language.

In Sprat's analysis, there was no other field that had suffered more from an excess of rhetoric than natural philosophy. If the Royal Society were to avoid the kind of obscure and extravagant language that was favoured among the Scholastics and Alchemists, they would have to apply some extremely harsh measures.[72]

> They have therefore been most rigorous in putting in execution, the only Remedy, that can be found for this *extravagance*: and that has been, a constant Resolution, to reject all the amplifications, digressions, and swellings of style: to return back to the primitive purity, and shortness, when men deliver'd so many *things*, almost in an equal number of *words*. They have exacted from all their members, a close, naked, natural way of speaking; positive expressions; clear senses; a native easiness: bringing all things as near the Mathematical plainness, as they can: and preferring the language of Artizans, Countrymen, and Merchants, before that, of Wits, or Scholars.[73]

Previous critics have particularly been concerned with two aspects of this statement outlining the Royal Society's attempted reform of language. First, the question has been raised as to its wider implications: did the Royal Society really make a serious effort to ban all methods of rhetorical amplification? If so, what consequences did this have on the literature of the period? The other aspect that has received much attention from modern scholars, concerns the extent to which the Society's linguistic "program" deviates from rhetorical convention.

According to Thomas M. Conley, even if the Society did not set out to reform any other kinds of writing besides the scientific, the statements on language

71 *The History*, 112–113.

72 *The History*, p. 113. On the Scholastic style, see *The History*, 15–16. On p. 37 of the same work, Sprat offers the following characterisation of "The Chymists," i.e. the alchymists: "Their writers involve them in such darkness; that I scarce know, which was the greatest task, to understand their meaning, or to effect it."

73 *The History*, 113 (italics in the original).

that were presented in Sprat's *History* did in fact have an immediate effect on writers such as the poet-playwright John Dryden (1631–1700).[74] In the view of Wilbur S. Howell, a dramatic shift had taken place in the history of rhetoric in the period between the publication of George Puttenham's *Arte of English Poesie* (1589) and Thomas Sprat's *History of the Royal Society*: even if one takes into account that the one devises "a rhetoric for the poet" and the other "a rhetoric for the scientist," Sprat's *History* represents a watershed moment when "the rhetoric of persuasion" is being substituted with "a new rhetoric of exposition."[75]

During the last 30–40 years, the concept both of the Scientific Revolution and of distinctive literary "periods" has been contested, and it no longer appears controversial to treat both Thomas Sprat and John Dryden as representatives of, not a distinct shift of paradigm, but a complex time of upheaval.[76] Moreover, even if one assumes that the Royal Society's views on language did in fact have a strong impact on other fields besides the scientific, how are we to prove that this was an automatic or necessary development? It is not hard to find examples in which the poet John Dryden (who was an early member of the Society) pays respect to the linguistic ideals that are voiced in Sprat's *History*, but Dryden's works also offer rich evidence that he was very much concerned with finding methods of pleasing and moving the audience, thus obeying rhetorical convention rather than any specific demands made by a contemporary institution.[77]

As for Thomas Sprat, he is no longer regarded only as a representative of a "new" rhetoric in the tradition of Wilbur S. Howell—it is also quite possible to

74 Conley, *Rhetoric in the European Tradition*, cited in n. 68 above, 169–170.

75 Howell, *Logic and Rhetoric in England, 1500–1700*, cited in n. 32 above, 388–390. See also Howell, *Eighteenth-Century British Logic and Rhetoric*, cited in n. 32 above, 482–487. In Howell's view, the origins of "the new rhetoric" can be traced back to the mid-1640s, when John Wilkins and his future co-founders of the Royal Society first gathered together at Oxford, and when Wilkins published his rhetorical work, *Ecclesiastes, or, A Discourse Concerning the Gift of Preaching As It Fals under the Rules of Art* (1646). See Howell, *Eighteenth-Century British Logic and Rhetoric*, 451–462.

76 The concept the Scientific Revolution came under attack in T.S. Kuhn, *The Structure of Scientific Revolutions* (Chicago: University of Chicago Press, 1962). For an overview of the arguments raised against the concept of distinct, historical periods, see D. Perkins, *Is Literary History Possible?* (Baltimore: Johns Hopkins University Press, 1992), 1–27.

77 See L. Feder, "John Dryden's Use of Classical Rhetoric," in *Essential Articles for the Study of John Dryden* ed. Swedenberg (London: Frank Cass and Company, 1966), 493–518; T. Skouen, *Passion and Persuasion. John Dryden's* The Hind and the Panther (Saarbrücken: VDM Verlag Dr. Müller, 2009).

see him through Ciceronian glasses, as the present study argues.[78] Furthermore, most scholars now tend to agree that there was no organized campaign against rhetoric, a change of view that is mainly due to Brian Vickers' effective refutation in the mid-1980s of the influential claim made by the American literary scholar Richard Foster Jones (1886–1965), that the Royal Society and, before that, Francis Bacon, were generally hostile to rhetoric.[79]

As we have seen, Thomas Sprat describes Bacon's style as distinctively imaginative and metaphorical, and Sprat also likes to express himself in metaphor. A brilliant example is when Sprat attempts to illustrate the Scholastics' outrageously figurative style by extravagantly comparing it to the feathers adorning the Native Americans: "like the Indians, [the Schole-men] onely express'd a wonderful artifice, in the ordering of the same feathers into a thousand varieties of figures."[80] Apparently, Sprat can see no good reason why he should not use the kind of imagery he warns so strongly against in other parts of his book, the reason being that these warnings are meant to apply to scientific writings, and not to poems or prose narratives such as *The History of the Royal Society*.

New Science, New Rhetoric?

When it comes down to it, the scientists of the age do not appear to have obeyed the rule not to use figurative language either.[81] As for the demand that all members of the Royal Society should strive to avoid digressions

---

78 The study is based on a paper held at *Science and Rhetoric*, The Norwegian Academy of Science and Letters' 150th Anniversary Symposium 2007. The Ciceronian basis of Sprat's argument has also been briefly mentioned in J. Richards, *Rhetoric* (London: Routledge, 2008), 7, 72–73.

79 Vickers, "The Royal Society and English Prose Style. A Reassessment," cited in n. 59 above. Vickers' critique centered on two studies in particular, namely the collection of essays by R.F. Jones entitled *The Seventeenth Century. Studies in The History of English Thought and Literature from Bacon to Pope* (Stanford: Stanford University Press, 1951) and R.F. Jones, *Ancients and Moderns. A Study of the Rise of the Scientific Movement in Seventeenth-Century England* (St. Louis: Washington University Press, 1961). For a helpful survey of the linguistic views of the Royal Society, see R. Nate, "Rhetoric in the Early Royal Society," in *Rhetorica Movet. Studies in Historical and Modern Rhetoric in Honour of Heinrich F. Plett*, eds. T.O. Sloane and P.L. Oesterreich (Leiden: Brill, 1999), 215–231. For a more detailed account, see W. Hüllen, *"Their Manner of Discourse". Nachdenken Über Sprache Im Umkreis Der Royal Society* (Tübingen: Gunter Narr, 1989).

80 *The History*, 15–16.

81 See J. Fahnestock, *Rhetorical Figures in Science* (New York: Oxford University Press, 1999).

and state their matter as briefly as they could, this goal was not easily joined with their parallel wish to be able to document each experiment in such detail that it would be possible for others to repeat them in order to test the results.[82]

One might ask, too, whether the Society's linguistic ideals represented anything new. It hardly seems progressive to want "to return back" to a time "when men deliver'd so many things, almost in an equal number of words."[83] The reference is most likely to the isomorphic vocabulary of Eden, or else to the universal language that was spoken before the fall of Babel.[84] This dream of recreating the original language was very much alive in the seventeenth century, and in his monumental work, *An Essay Towards a Real Character and a Philosophical Language* (which appeared about the same time as Sprat's *History*), the Royal Society's founding member, John Wilkins, asserts that "every word ought in strictness to have but one proper sense and acception, to prevent equivocalness."[85] Even though Wilkins went to great lengths in his *Essay Towards a Real Character* to devise a new, universal language, the idea that one might recreate the pre-Babelian form of communication was nothing new, because similar plans had been launched at various times in different cultures.[86] As regards the call for an isomorphic correspondence between things and words, this was not something that originated in the Royal Society, nor was it exclusively linked with the renewed interest among seventeenth-century thinkers in the categorization of nature.

In England there was a real need for reducing the equivocal quality of the vernacular, which had been—and still was—characterized by much regional and individual variation.[87] This fact sets the linguistic views of the Royal Society into perspective. After all, as I mentioned earlier, its founding members were not just natural philosophers, but some were also distinguished linguists. While John Wallis made a lasting contribution to the new vernacular grammar, John Wilkins not only invented a new kind of language to be used in some

82 Cf. Shapin, *The Scientific Revolution*, cited in n. 6 above, 107–108.

83 *The History*, 113 (italics in the original).

84 *Genesis* 2.20, 11.1–9.

85 Quoted from the facsimile edition of J. Wilkins, *An Essay Towards a Real Character and a Philosophical Language, 1668, English Linguistics 1500–1800*, No. 119, ed. R.C. Alston (Menston: Scolar Press, 1968), 318.

86 See U. Eco, *The Search for the Perfect Language* (Oxford: Blackwell, 1995).

87 See M. Görlach, "Regional and Social Variation," in *The Cambridge History of the English Language*, vol. 3: 1476–1776 ed. R. Lass (Cambridge: Cambridge University Press, 1999), 459–538.

utopian future, but also produced a solid alphabetical dictionary based on such words as were actually in use among his contemporaries.[88]

In the age of Sprat, the process of developing a standardized grammar and pronunciation was still only in its beginning, and the English language still met with some competition from French and Latin.[89] The English tongue had yet to become a *lingua franca*, as one can see from the reception of Sprat's *History* on the Continent. Just as is the case with most scientists and scholars today, the Royal Society was eager to attract potential partners outside its own country, but unlike today, English was not the best language to use when applying for funding or communicating with fellow academics across borders. In fact, there were so few "out there" who could read Sprat's "project description" without any trouble, that within just one year of the publication of *The History of the Royal Society* a French translation was made available. As for the planned Latin version, this never materialized.[90] Although Latin would continue to play a significant role as the language of the learned for another century yet, it was losing ground.[91] This development, too, can be illustrated by reference to Sprat's *History of the Royal Society*.

Both in Sprat's book and in a later work by the same title, Thomas Birch's more reliable *History of the Royal Society* in four volumes dating from 1756–1757, there is a transcript of the Royal Charter that the Royal Society had received in 1662.[92] But whereas Birch quotes the original Latin text in full, Sprat

88 The dictionary is included at the end of the facsimile edition of J. Wilkins, *An Essay Towards a Real Character*, cited in n. 85 above.

89 The point about competition from French and Latin is made by G. Knowles, *A Cultural History of the English Language* (London: Arnold, 1997), 107. The process of standardization is outlined by V. Salmon, "Orthography and Punctuation," in *The Cambridge History of the English Language*, vol. 3, ed. Lass, cited in n. 87 above, 13–55. For a very brief summary, see M. Görlach, *Aspects of the History of English*, Anglistische Forschungen, vol. 260 (Heidelberg: Universitätsverlag C. Winter, 1999), 2–3.

90 See *Dictionary of Scientific Biography*, vol. 12, cited in n. 1 above, 583, where one can also read that *The History* sold well in Sprat's home country and that he received much praise for his eloquent prose. According to Cope and Jones, the idea of producing a Latin translation came from John Wilkins, see *The History*, x, n. 5.

91 It has been suggested that the Latin tradition had slowly begun to wane in England from the 1650s onwards, see J.W. Binns, *Intellectual Culture in Elizabethan and Jacobean England*, *Arca. Classical and Medieval Texts, Papers and Monographs*, vol. 24 (Leeds: Francis Cairns, 1990), 392.

92 Compare *The History*, 134–143, with T. Birch (1705–1766), *The History of the Royal Society of London*, vol. 1 (London: A. Millar, 1756–1757), 88–96. As explained in n. 10 above, Sprat's paraphrase actually contains elements both from the first Charter of 1662 and the second Charter of 1663.

gives a paraphrase in English. In Birch's case, his decision to include the original text shows how the learned reader of the mid-eighteenth century was still able to read Latin. By comparison, Sprat's choice to translate the King's words into the vernacular a century before appears as demonstratively inclusive: his work is not only addressed to the learned or to those who had received a Latin education.

In the late seventeenth century, there was an increasing concern with extending the use of the vernacular to all fields of society and this aim is clearly signaled by the Royal Society's decision to abandon the language of "wits" and "scholars" (that is, the language associated with courtly culture and bookish studies), in favor of such words as are used by "Artizans, Countrymen, and Merchants."[93] The Royal Society's call for clarity and plain expression should therefore not be regarded as something that pertained only to a certain kind of learned writing, but it was part of a national endeavor to stabilize the language. This point was actually made (although only very briefly) by Richard Foster Jones in *The Triumph of the English Language*, but it has not since received due attention.[94]

The process both of stabilizing and authorizing the mother tongue (in opposition with Latin) must also have had a severe impact on rhetoric. It is not hard to imagine that the persistent efforts to describe and define the vernacular standard in grammars and dictionaries drew attention away from the rhetorical discipline. Whereas the new, vernacular grammar was to play an important role in modern society, rhetoric would be linked with the ancient, Latin tradition. From this perspective it was the grammarians – not the scientists—that were rhetoric's worst enemies.

So far, we may conclude that the Royal Society's views on language were partly progressive and partly regressive. While the emphasis on the vernacular was forward-looking, the idea of recreating a previous linguistic state seems quite the opposite. Sprat's ideas of a "primitive purity, and shortness" and "positive expressions; clear senses," as opposed to "the amplifications, digressions, and swellings of style," also recall the ancient contrasts between *perspicuitas* and *obscuritas*, *copia* and *brevitas*, as well as the classical rhetoricians' emphasis on pure or idiomatically and grammatically correct expression (*latinitas*, *sermo purus*).[95] By placing definite statements before figurative expressions, Sprat furthermore observes the traditional division between logical demonstration

93 *The History*, 113.

94 R.F. Jones, *The Triumph of the English Language. A Survey of Opinions Concerning the Vernacular from the Introduction of Printing to the Restoration* (Stanford: Stanford University Press, 1953), 311.

95 *The History*, 113. The Latin terms do not appear in Sprat's work.

and persuasive argumentation, or even between the arts of logic and rhetoric in the Latin curriculum. Finally, his ideal of a "naked, natural way of speaking" also points back in time, in so far as Plato and Aristotle are the best models he can think of.[96]

The same ambiguous attitude—partly progressive and partly regressive—also characterizes the Royal Society's motto, *nullius in verba*, which has alternately been translated by the Society as "nothing in words" or "take nobody's word for it."[97] The motto thus reflects the members' express wish to study nature with their own eyes, independently of such dogmatic truths as they would find in books.[98] At the same time, the motto curiously undermines such desire for independence, since the motto was based on Horace's *Epistles*.[99] In addition, the translation "nothing in words" does not seem a wise choice for an institution that was eager to spread the news about its recent discoveries and about its very existence in books like *The History of the Royal Society* and in journals like the *Philosophical Transactions* series, which still continues today. The choice of a Latin motto too shows that the classical language had not yet lost its power.

In the final part of this study, I would like to draw attention to the no less ambiguous attitude underlying Sprat's eloquent defense of science. I am not primarily thinking of the fact that he uses rhetorical devices while at the same time arguing against the use of such devices, since the proposed reform of language was not intended to apply to the kind of prose Sprat was producing anyway. Instead, I would like to emphasize how Sprat is also building his argument on ideas deriving from Ciceronian rhetoric, a strategy that might work in two directions, either rejecting or confirming the authority of the inherited knowledge.

Eloquence versus Wisdom

Depending on the context, Sprat uses the word eloquence in both positive and negative senses. When he talks about Francis Bacon's argument on behalf of

96 Quotation taken from *The History*, 113. *The History*, 16, where Sprat says of Plato and Aristotle that "they always strove to be easie, naturall, and unaffected."

97 In 2008, the preferred translation on the Royal Society website was "nothing in words." (http://royalsociety.org/page.asp?id=6186 accessed Dec. 8, 2008). In 2009, the Royal Society updated the presentation of their motto, substituting "nothing in words" with "take nobody's word for it." http://royalsociety.org/Nullius-in-verba (accessed Nov. 27, 2009).

98 One should also consider how Sprat describes the nature of philological studies in *The History*, 24–25.

99 The Royal Society refers to *Epistles* I.i.14: *Nullius addictus iurare in verba magistri*. http://royalsociety.org/Nullius-in-verba (accessed Nov. 27, 2009).

experimental science, Bacon's eloquence is undeniably a good thing, but when he calls to mind the way in which words can create political turmoil, he is inclined to think it would be best if eloquence were to be banished from "all civil societies," as something that can have a fatal effect on "peace and good manners."[100]

This thought represents a clear breach with the rhetorical tradition, which rather holds that eloquence is the driving force of civilization. According to Cicero, the power of eloquence was what had first led mankind out of "its brutish existence in the wilderness."[101] This myth about the civilizing power of language was based on a basic distinction between beasts and men in terms of reason and speech: man is a rational animal, and his capacity for reason reveals itself through his speech.[102] As Sprat's contemporary, the philosopher Thomas Hobbes (1588–1679), maintained, if it had not been for the invention of speech, man would have been no more capable of forming societies than "lions, bears, and wolves."[103] Sprat actually plays upon such commonplace assimilation of reason with speech when he describes King Charles I (who reigned from 1625 until he was beheaded in 1649) as "an inimitable master, in reason and eloquence."[104]

It may seem paradoxical that Sprat should praise the King's eloquence, if he believes that eloquence has no place in a civil society. But the key to understanding this apparent inconsistency lies in Sprat's remark that eloquence can be a dangerous weapon if it falls into the wrong hands.[105] Such warnings against the possible abuse of eloquence were as old as rhetoric itself, but whereas Cicero and Quintilian had sought to eliminate this fear by claiming that only the good man could become a good speaker, and that the best speakers combine eloquence with wisdom, Sprat rather claims that this original bond between eloquence and wisdom has been broken.[106] For, as he argues, in

---

100 Quoted from *The History*, 111. Cf. *The History*, 4–5.

101 *De or.*, I. 33. Trans. E.W. Sutton and H. Rackham, *The Loeb Classical Library* 348 (Cambridge, MA: Harvard University Press, 1942).

102 *De or.*, I. 32; Quintilian, *Inst. or.*, VII. iii. 15.

103 T. Hobbes, *Leviathan* [*1651*], *Oxford World Classics*, ed. J.C.A. Gaskin (Oxford: Oxford University Press, 1998), 20.

104 *The History*, 152.

105 *The History*, 111.

106 Cicero, *De inv.*, I. 1 and *De or.*, III. 55–82; Quintilian, *Inst. or.*, I. Pr. 9–11. See also the dedication in the facsimile edition of H. Peacham, *The Garden of Eloquence, 1577*, *English Linguistics 1500–1800*, No. 267 (Menston: Scolar Press, 1971), A2v: "wisdom doe require the light of Eloquence, and Eloquence the fertillity of Wysedome." As Amund Børdahl (University of Bergen) pointed out to me at The Fifth Conference of the Nordic Network

his own day and age rhetorical figures were mainly used in defiance of reason, to stir the rebellious passions, the terrible result of which had been the outbreak of the Civil War. "[T]hose dreadful revolutions…cannot be beheld on paper, without horror," Sprat writes, adding a soothing observation that when the King returned to power, this had put an end to the terror.[107]

Considering the immediate political history, one can understand why Sprat cannot embrace eloquence in the same way as, say, the Renaissance rhetorician Thomas Wilson had done in his preface to *The Arte of Rhetorique* (1553), where Wilson says that the one who excels in those faculties "wherin men do excell beastes" deserves to be praised as being "halfe a God."[108] Because when the wicked speaker—represented by the wartime leader, Oliver Cromwell—gained victory over the wise speaker (King Charles I), the premises were changed. The gift of eloquence no longer brought out the best in man.

As a writer, Sprat is therefore bound to reconsider the accepted truths deriving from the standard textbooks in his field in much the same way as the natural philosophers have done in their respective fields. In both cases, the handed-down knowledge is reviewed through first-hand experience and observation. At the same time, neither Sprat nor the scientists end up rejecting all they have learnt before. Whereas the natural philosophers are still tied to the Aristotelian world picture, Sprat continues to conceive of society or "civil business" and "humane affairs" in Ciceronian terms.[109] Following Cicero, he maintains that man is invested with a distinctive, rational drive that makes him capable of acting for the common good, but Sprat differs from Cicero in that this rational drive or power is not thought to be expressed in words, but rather through the practice of science.

By attributing the same fundamental function to scientific learning as had traditionally been attributed to eloquence, Sprat not only affirms the power of science, but he also gives an idea that the English Royal Society is at the head of human progress. Throughout his work, Sprat is eager to tell how England has become the world's leading nation in commerce, learning, and technology.[110]

---

for the History of Rhetoric in 2009, Sprat would also have found the basis for his argument concerning the divorce between wisdom and eloquence in classical rhetoric: in *De or.*, III. 72, Cicero holds Socrates responsible for the split between philosophy and oratory.

107 *The History*, 58.

108 Quoted from the facsimile edition of T. Wilson, *Arte of Rhetorique*, *The English Experience*, vol. 206 (Amsterdam: Da Capo Press, 1969), A4 (misprinted as A3). Cf. *De or.*, I. 33; Peacham, *The Garden of Eloquence, 1577*, cited in n. 106 above, A2.

109 Quotations taken from *The History*, 56.

110 *The History*, 86–88, 114–115, 150–153.

This has been because the English are naturally suited to rule the world: not only do they have the right temperament and the perfect geographic location (being situated in the passageway between North and South), but they are also prepared to work hard, in both a manual and mental sense.[111] To the extent that there have been some hindrances to the country's successful development, this has either been because of political and religious conflict or because the pursuit of knowledge has been limited by the study of ancient texts.[112] But now, Sprat says, all this is a thing of the past: "now not only the eyes of men, but their hands are open, and prepar'd to labour."[113] It is fascinating to observe how Sprat employs the image of the open hand—not in its conventional sense, to describe the art of rhetoric, but to serve his argument that the future will depend on actions rather than words.[114]

Already in the age of Elizabeth I it had become clear what a great future was ahead: "commerce was establish'd, and navigation advanc'd."[115] But the time had not yet come to build a formal institution for the promotion of scientific experimentation and innovation.[116] During the reign of James I, the time was still not ripe, and thus the scientific society which Francis Bacon described in his utopia titled *The New Atlantis* (1627) was just that—a utopia.[117] Then, when Charles I came to the throne, there was a change of attitude, Charles not only being a master of "reason and eloquence" but also of all sorts of practical skills.[118] But the Royal Society was not founded until the age of Charles II, who had not only been acting as the patron of science, but had even taken an active part in the Society's meetings and "assisted with his own hands," Sprat claims, "at the performing of many of [the Society's] experiments."[119] This seems greatly exaggerated. Apparently, the Society was quite offended that Charles did not seem to take much interest in their proceedings, and on one of those few occasions when he is reported to have been present, he is said to have burst out laughing.[120]

---

111 *The History*, 114–115. See also 420.

112 *The History*, 150–152.

113 *The History*, 152.

114 On the commonplace association of the open hand with the art of public speaking, see Howell, *Logic and Rhetoric in England, 1500–1700*, cited in n. 32 above, esp. pp. 14–15.

115 *The History*, 151.

116 *The History*, 151.

117 *The History*, 151–152.

118 *The History*, 151.

119 *The History*, 133.

120 See *The History*, 133, and Cope and Jones's note on *The History*, 133 (*The History*, Notes, 14).

Evidently, Sprat is less concerned with telling the truth than with getting his message across that it is time to stop talking and start working.[121] Because the thinkers and innovators of the Royal Society no longer need to be slaves to the past, they can concentrate on developing new methods and insights "to the uses of humane society."[122] By focusing only on "material things," there is little chance the Royal Society will fall into the trap of "talking, instead of working," which is what happens whenever one treats of politics, rhetoric, ethics, or any other subject that depends on judgment and opinion.[123] Whereas the arts of "Politicks, Morality, and Oratory" may have been important in the early stages of Western civilization, they no longer appear useful.[124] What the world needs now, is to gain more solid knowledge about "the works of nature," the working of the body (a "natural engine"), and the great potential that lies in "the arts of [men's] hands."[125] Sprat presupposes that man does not primarily reveal his capacity for rational and civil behavior through verbal communication, but by developing new skills and knowledge: "For methinks there is an agreement, between the growth of learning, and of civil government," he says.[126]

Science, not eloquence, is the driving force of civilization, and this is Sprat's most powerful argument when towards the end of his work he appeals for moral and financial support. If the Royal Society of London does not get the funding needed to proceed with its national, scientific endeavor, the whole world will suffer.[127] By supporting the Royal Society, one will not only be serving this particular institution, but the whole of humanity:

> if…our Nation shall lay hold of this opportunity, to deserve the applause of Mankind, the force of this Example will be irresistibly prævalent in all Countries round about us; the State of Christendom will soon obtain a new face; while this Halcyon Knowledge is breeding, all tempests will cease: the oppositions and contentious wranglings of Science falsly so call'd, will soon vanish away: the peaceable calmness of [men's] Judgments, will have admirable influence on their Manners; the sincerity

121 Cf. *The History*, 423. See also 62, where Sprat contrasts "a glorious pomp of words" with "solid practice…and unanswerable arguments of real productions."

122 Quoted from *The History*, 83. Cf. *The History*, 29.

123 *The History*, 82, 118.

124 *The History*, 82, 29 (p. 82).

125 *The History*, 82–83.

126 *The History*, 29.

127 *The History*, 437.

> of their Understandings will appear in their Actions; their Opinions will be less violent and dogmatical, but more certain; they will only be Gods one to another, and not Wolves; the value of their Arts will be esteem'd by the great things they perform, and not by those they speak.[128]

Here, Sprat makes it appear that by promoting the new science, the English nation will not only set an example that is worthy of applause, but it will represent a civilizing force that will change how people think and behave forever. For when they gain true knowledge, they will be less inclined to fight: they will behave as rational men or as gods towards one another, instead of wolves. And because they develop all sorts of useful arts and technologies, these men will be remembered – not for what they *said*, but for what they *did* for the good of mankind.

This concluding argument in *The History of the Royal Society* represents a far more dramatic breach with the rhetorical tradition than those surly remarks about eloquence for which Sprat is notorious. To the extent that Sprat's work contributed to the decline of rhetoric, it was perhaps not so much because of what he said about eloquence, but because of how he described the role of science. At the same time, I would argue that Sprat's humanistic argument in favor of science also represents a continuation of the rhetorical tradition.

Consider first, how Sprat describes his own education as a process of indoctrination, whereby one learns to appreciate the art of rhetoric without question.[129] We may ask whether it was even possible for a writer at this time to conceive of civilization without any reference to the classical-rhetorical terminology. At school, emphasis would be on how to present variations on a common theme, which is part of what Sprat is doing when he adapts the Ciceronian material to his own persuasive purpose. Yet, Sprat does more than to recycle the existing material, because his handling of Cicero is not just a matter of *copia* or variation but even more of *translatio*: the significance commonly attributed to eloquence is successfully transferred to another sphere, the sphere of science. It seems a rather daring experiment to try and substitute eloquence with science. By managing to pull this experiment through, Sprat not only proves his great rhetorical skill, but his published work also gives a demonstration of precisely the kind of method that the Royal Society was advocating, first trying the existing knowledge through experimentation and then communicating the results of the investigation in writing.

128 *The History*, 437–438.

129 *The History*, 112.

Sprat's manner of transferring and appropriating the Ciceronian ideology also serves to illustrate the point he is making on behalf of natural science, namely that an inherited truth should not automatically be passed on to future generations, but that it needs to be reviewed in the light of recent developments.[130] Sprat cannot possibly repeat what Cicero said about civilization, if the commonplace claim about wisdom and eloquence being two sides of the same coin no longer rings true to Sprat's audience. In addition, the theory and practice of rhetoric will not be of much help in a society that most of all needs to develop new sciences and technologies. In order to outline the future state of mankind, Sprat will therefore have to dispose of much of the classical-rhetorical material. But the idea that man has a rational drive inside that ensures good progress still seems to fit, so this idea he decides to keep. In such respect, Sprat contributes to the preservation of a fundamental piece of the humanist heritage at a time when the Latin-based culture was on the wane. As we may remember, Sprat compares the advancement of knowledge to an adventurous sea voyage, and, to borrow Sprat's metaphor, it appears he has rescued whatever he found worth rescuing from the sinking flagship named rhetoric.[131]

Time for Action

Looking back, one may ask whether Sprat's rescue operation was a success. Can the humanist ideology survive in the modern world of science and technology? If one visits the Royal Society's homepage on the Internet, one can see that the humanistic arguments in favor of science are still very much in use. Speaking through the world wide web, the Society is just as eager to tell the whole wide world that it is "a Fellowship of the world's most distinguished scientists, which promotes the advancement of science and its use for the benefit of humanity and the good of the planet."[132] With reference to Francis Bacon, the Society's former President Martin Rees (2005–2010) maintains that "scientific discovery should be driven not just by the quest for intellectual enlightenment but also for the relief of man's estate."[133] This, he says, is now more true than ever, considering that "for the first time in its history, our planet's fate depends on human actions and human choices."[134] Arguing that science is

130 *The History*, 49–50.

131 *The History*, 109.

132 http://royalsociety.org/about-us (accessed Sept. 6, 2010).

133 http://royalsociety.org/Presidents-Message (accessed Sept. 6, 2010).

134 http://royalsociety.org/Presidents-Message (accessed Sept. 6, 2010).

"an integral part of our shared human culture," Lord Rees insists that "the astonishing advances that will come in the next decades must be used for the benefit of society worldwide."[135] The Royal Society's 350th anniversary celebration has thus not only been about honoring the past, but about creating a better future.

Just as in Sprat's *History*, the emphasis on the webpage is on what science can do—on science in action—and the same buzzword ("action") is used to appeal for philanthropic support: part of what everyone can do to beat climate change, is to make a donation so that the scientists can continue with their pivotal work.[136] As the current President of the Society says: "Our actions now will have a profound effect on the future."[137]

By partly promoting and partly appealing for support of both the natural and applied sciences in general and the Royal Society in particular, the website performs exactly the same function today as Sprat's *History* did in the 1660s. Clearly, the Society's ambitions have not been lowered as it works "to extend the revolution in science for the maximum benefit of the planet and its people."[138]

Today's website thus rests on an assumption that the progress of science and of man go hand in hand. According to the Royal Society, "science is a major driver of social and economic progress," and it should therefore be regarded as deeply embedded "in culture and human progress."[139] It depends on the eye of the beholder if this argument means that rhetoric has been outmaneuvered by science as the prime motivator of civilization, or if by adopting the Ciceronian take on humanity, the advocators of science rather testify to rhetoric's continued influence.

---

135 http://royalsociety.org/Presidents-Message (accessed Sept. 6, 2010).

136 http://royalsociety.org/support-us, http://royalsociety.org/enterprisefund (accessed Sept. 6, 2010). Similar appeals to induce potential benefactors to open their purses can be found in *The History*, 125, 437.

137 http://royalsociety.org/Presidents-Message (accessed Sept. 6, 2010).

138 http://royalsociety.org/campaign (accessed Sept. 6, 2010). See also http://royalsociety.org/Strategic-Priorities (accessed Sept. 6, 2010).

139 http://royalsociety.org/Strategic-Priorities and http://royalsociety.org/Inspire-interest-in-the-joy-wonder-and-excitement-of-scientific-discovery (accessed Sept. 6, 2010).

Further Reading

Our list of material for further reading is by no means comprehensive. Rather, our intent is to supplement the lists of works cited in the Introduction and individual articles with additional sources that reinforce key themes.

Agassi, Joseph, *The Very Idea of Modern Science. Francis Bacon and Robert Boyle* (Heidelberg: Springer, 2013).

Anstey, Peter, and Alberto Vanzo, "The Origins of Early Modern Experimental Philosophy," *Intellectual History Review* 22, 4 (2012), 499–518.

Atkinson, Dwight, *Scientific Discourse in Sociohistorical Context. The Philosophical Transactions of the Royal Society of London, 1675–1975* (New York: Routledge, 1998).

Beeley, Philip, "Auf der Suche nach neuen Wegen. Die Selbstdarstellung der Leopoldina und der Royal Society in ihrer Korrespondenz," *Acta Historica Leopoldina* 49 (2008), 305–326.

Bektas, M. Yakup, and Maurice Crosland, "The Copley Medal. The Establishment of a Reward System in the Royal Society, 1731–1839," *Notes and Records of the Royal Society of London* 46, 1 (1992), 43–76.

Ben-Chaim, Michael, *Experimental Philosophy and the Birth of Empirical Science. Boyle, Locke and Newton* (Burlington: Ashgate, 2004).

Bennett, James, "Shopping for Instruments in Paris and London," in *Merchants and Marvels. Commerce, Science, and Art in Early Modern Europe*, eds. Pamela Smith and Paula Findlen (London: Routledge, 2002), 370–395.

Bensaude-Vincent, Bernadette, and Christine Blondel (eds.), *Science and Spectacle in the European Enlightenment* (Burlington: Ashgate, 2008).

Blair, Ann, *Too Much to Know. Managing Scholarly Information before the Modern Age* (New Haven: Yale University Press, 2010).

Boschiero, Luciano, *Experiment and Natural Philosophy in Seventeenth-Century Tuscany. The History of the Accademia Del Cimento* (Dordrecht: Springer, 2007).

Briggs, Robin, "Religion and Natural Philosophy," in *Early Modern Europe. An Oxford History*, ed. Euan Cameron (Oxford: Oxford University Press, 1999), 171–205.

Burns, William A., *An Age of Wonders. Prodigies, Politics and Providence in England 1657–1727* (Manchester: Manchester University Press, 2010).

Byden, D.J., and Simms, D.L., "Spectacles Improved to Perfection and Approved of by the Royal Society," *Annals of Science* 50, 1 (1993), 1–32.

Campbell, Mary Baine, *Wonder and Science. Imagining Worlds in Early Modern Europe* (Ithaca: Cornell University Press, 1999).

Carey, Daniel, "Compiling Nature's History. Travellers and Travel Narratives in the Early Royal Society," *Annals of Science* 54, 3 (1997), 269–292.

Cook, Harold, "Moving About and Finding Things Out. Economies and Sciences in the Period of the Scientific Revolution," *Osiris* 27 (2012), 101–132.

Cregan, Kate, *The Theater of the Body. Staging Death and Embodying Life in Early Modern London* (Turnhout: Brepols, 2009).

da Costa, P. Fontes, "The Culture of Curiosity at The Royal Society in the First Half of the Eighteenth Century," *Notes and Records of the Royal Society of London* 56, 2 (2002), 147–166.

Dear, Peter, "Mysteries of State, Mysteries of Nature. Authority, Knowledge and Expertise in the Seventeenth Century," in *States of Knowledge. The Co-Production of Science and Social Order*, ed. Sheila Jasanoff (London: Routledge, 2004), 206–224.

——, "A Philosophical Duchess. Understanding Margaret Cavendish and the Royal Society," in *Science, Literature and Rhetoric in Early Modern England*, eds. Juliet Cummins and David Burchell (Burlington: Ashgate, 2007), 125–143.

——, "From Truth to Disinterestedness in the Seventeenth Century," *Social Studies of Science* 22, 4 (1992), 619–631.

Dickson, Donald, "Afterword. 'Philosophicall and Mathematico-Mechanical King. Charles II and the Royal Society,'" in *The Tessera of Antilia. Utopian Brotherhoods and Secret Societies in the Early Seventeenth Century* (Boston: Brill, 1998), 237–245.

Freddi, Maria, Barbara Korte, and Josef Schmied, "Developments and Trends in the Rhetoric of Science," *European Journal of English Studies* 17, 3 (2013), 221–234.

Garber, Daniel, and Michael Ayers (eds.), *The Cambridge History of Seventeenth-Century Philosophy*, 2 vols. (Cambridge: Cambridge University Press, 1998).

Gieryn, Thomas, "Distancing Science from Religion in Seventeenth-Century England," *Isis* 79, 4 (1988), 582–593.

Grafton, Anthony, *Defenders of the Text. The Traditions of Scholarship in an Age of Science, 1450–1800* (Cambridge: Harvard University Press, 1991).

Hall, Marie Boas, *Henry Oldenburg. Shaping the Royal Society* (Oxford: Oxford University Press, 2002).

Hanson, Craig Ashley, *The English Virtuoso. Art, Medicine, and Antiquarianism in the Age of Empiricism* (Chicago: University of Chicago Press, 2009).

Harkness, Deborah, *The Jewel House. Elizabethan London and the Scientific Revolution* (New Haven: Yale University Press, 2007).

Harrison, Peter, "Religion and the Early Royal Society," *Science & Christian Belief* 22, 1 (2010), 3–22.

——, "Religion, the Royal Society, and the Rise of Science," *Theology and Science* 6, 3 (2008), 255–271.

Heyd, Michael, "The New Experimental Philosophy. A Manifestation of 'Enthusiasm' or an Antidote to It?" *Minerva* 25, 4 (1987), 423–440.

Hunter, Michael, "The Great Experiment," *History Today* 60, 11 (2010), 34–40.

Iliffe, Robert, "Foreign Bodies. Travel, Empire and the Early Royal Society of London, Part I. Englishmen on Tour," *Canadian Journal of History* 33, 3 (1998), 357–385.

——, "Foreign Bodies. Travel, Empire and the Early Royal Society of London, Part II. The Land of Experimental Knowledge," *Canadian Journal of History* 34, 1 (1999), 23–50.

Israel, Jonathan I., *Enlightenment Contested. Philosophy, Modernity, and the Emancipation of Man 1670–1752* (Oxford: Oxford University Press, 2006).

Jardine, Lisa, "Monuments and Microscopes. Scientific Thinking on a Grand Scale in the Early Royal Society," *Notes and Records of the Royal Society of London* 55, 2 (2001), 289–308.

Jarvis, J. Ereck, "Thomas Sprat's 'Mixt Assembly.' Association and Authority in the History of the Royal Society," *Restoration* 37, 2 (2013), 55–77.

Johns, Adrian, "Identity, Practice, and Trust in Early Modern Natural Philosophy," *The Historical Journal* 42, 4 (1999), 1125–1145.

Killeen, Kevin, and Peter Forshaw (eds.), *The Word and the World. Biblical Exegesis and Early Modern Science* (New York: Palgrave, 2007).

Krips, Henry, "Ideology, Rhetoric, and Boyle's New Experiments," *Science in Context* 7, 1 (1994), 53–64.

Levitin, Dmitri, "The Experimentalist as Humanist. Robert Boyle on the History of Philosophy," *Annals of Science* 71, 2 (2014), 149–182.

Lynch, William, *Solomon's Child. Methods in the Early Royal Society of London* (Stanford: Stanford University Press, 2001).

Malcolmson, Cristina, *Studies of Skin Color in the Early Royal Society. Boyle, Cavendish, Swift* (Burlington: Ashgate, 2013).

Martin, Catherine Gimelli, "Rewriting the Revolution. Milton, Bacon, and the Royal Society Rhetoricians," in *Science, Literature and Rhetoric in Early Modern England*, eds. Juliet Cummins and David Burchell (Burlington: Ashgate, 2007), 97–124.

Meli, Domenico Bertoloni, "Early Modern Experimentation on Live Animals," *Journal of the History of Biology* 46, 2 (2013), 199–226.

Morgan, John, "Religious Conventions and Science in the Early Restoration. Reformation and 'Israel' in Thomas Sprat's *History of the Royal Society*," *British Journal for the History of Science* 42, 3 (2009), 321–344.

Mortimer, Sarah, and John Robertson (eds.), *The Intellectual Consequences of Religious Heterodoxy, 1600–1750* (Leiden: Brill, 2012).

Moss, Jean Dietz, *Novelties in the Heavens. Rhetoric and Science in the Copernican Controversy* (Chicago: University of Chicago Press, 1993).

Ochs, Kathleen, "The Royal Society of London's History of Trades Programme. An Early Episode in Applied Science," *Notes and Records of the Royal Society of London* 39, 2 (1985), 129–158.

Park, Katherine, and Lorraine Daston (eds.), *The Cambridge History of Science, vol. 3. Early Modern Science* (Cambridge: Cambridge University Press, 2006).

Pender, Stephen, and Nancy S. Struever (eds.), *Rhetoric and Medicine in Early Modern Europe* (Burlington: Ashgate, 2012).

Poole, William, "Francis Lodwick's Creation. Theology and Natural Philosophy in the Early Royal Society," *Journal of the History of Ideas* 66, 2 (2005), 245–263.

Porter, Roy, *Creation of the Modern World. The Untold Story of the British Enlightenment* (New York: Norton, 2001).

Rusnock, Andrea, "Correspondence Networks and the Royal Society, 1700–1750," *The British Journal for the History of Science* 32 (1999), 155–169.

Sarasohn, Lisa, "Thomas Hobbes and the Duke of Newcastle. A Study in the Mutuality of Patronage before the Establishment of the Royal Society," *Isis* 90, 4 (1999), 715–737.

Schneider, Pablo, "Scientific Images and the Mysteries of Power in the Era of Louis XIV," in *Philosophies of Technology. Francis Bacon and His Contemporaries*, 2 vols., vol. 1, eds. Claus Zittel, Gisela Engel, Romano Nanni, and Nicole C. Karafyllis (Leiden: Brill, 2008), 539–572.

Serjeantson, Richard, "Testimony. The Artless Proof," in *Renaissance Figures of Speech*, eds. Sylvia Adamson, Gavin Alexander, and Katrin Ettenhuber (Cambridge: Cambridge University Press, 2007), 179–194.

Sorrenson, Richard, "Towards a History of the Royal Society in the Eighteenth Century," *Notes and Records of the Royal Society of London* 50, 1 (1996), 29–46.

Stafford, Barbara, *Artful Science. Enlightenment Entertainment and the Eclipse of Visual Education* (Cambridge: MIT Press, 1994).

Strano, Giorgio, Stephen Johnston, Mara Miniati, and Alison Morrison-Low (eds.), *European Collections of Scientific Instruments, 1550–1750* (Leiden: Brill, 2009).

van Bunge, Wiep, "Dutch Cartesian Empiricism and the Advent of Newtonianism," *Studies in History and Philosophy of Science* 31 (2013), 89–104.

Vidal, Fernando, *The Sciences of the Soul. The Early Modern Origins of Psychology*, trans. Saskia Brown (Chicago: University of Chicago Press, 2011).

Walsh, Lydia, *Scientists as Prophets. A Rhetorical Genealogy* (New York: Oxford University Press, 2013).

Wardhaugh, Benjamin, *Music, Experiment and Mathematics in England*, 1653–1705 (Burlington: Ashgate, 2008).

Yeo, Richard, *Notebooks, English Virtuosi, and Early Modern Science* (Chicago: University of Chicago Press, 2014).

Index

www.ingramcontent.com/pod-product-compliance
Lightning Source LLC
LaVergne TN
LVHW010555300926
849162LV00002B/337